Urban Aerial Pesticide Spraying Campaigns

This book examines social processes that have contributed to growing pesticide use, with a particular focus on the role governments play in urban aerial pesticide spraying operations.

Beyond being applied to sparsely populated farmland, pesticides have been increasingly used in densely populated urban environments, and when faced with invasive species, governments have resorted to large-scale aerial pesticide spraying operations in urban areas. This book focuses on New Zealand's 2002–2004 pesticide campaign to eradicate the Painted Apple Moth, which is the largest operation of its kind in world history, whether we consider its duration (29 months), its scope (at its peak the spraying zone was 10,632 hectares/26,272 acres), the number of sprayings that were administered (the pesticide was administered on 60 different days), or the number of people exposed to the spraying (190,000+). This book provides an in-depth understanding of the social processes that contributed to the incursion, why the government sought to eradicate the moth through aerial pesticide spraying, the ideological strategies they used to build and maintain public support, and why those strategies were effective.

Urban Aerial Pesticide Spraying Campaigns will be of great interest to students and researchers of pesticides, environmental sociology, environmental history, environmental studies, political ecology, geography, medical sociology, and science and technology studies.

Manuel Vallée is a Senior Lecturer of Sociology in the School of Social Sciences at The University of Auckland, New Zealand.

Routledge Explorations in Environmental Studies

For more information about this series, please visit: www.routledge.com/
Routledge-Explorations-in-Environmental-Studies/book-series/REES

Urban Aerial Pesticide Spraying Campaigns

Government Disinformation, Industry Profits, and Public Harm

Manuel Vallée

First published 2023
by Routledge
4 Park Square, Milton Park, Abingdon, Oxon OX14 4RN

and by Routledge
605 Third Avenue, New York, NY 10158

Routledge is an imprint of the Taylor & Francis Group, an informa business

British Library Cataloguing-in-Publication Data
A catalogue record for this book is available from the British Library

Library of Congress Cataloging-in-Publication Data
Names: Vallée, Manuel, author.
Title: Urban aerial pesticide spraying campaigns : government
disinformation, industry profits, and public harm / Manuel Vallée.
Description: First edition. | Abingdon, Oxon ; New York, NY :
Routledge, 2023. |
Series: Routledge explorations in environmental studies |
Includes bibliographical references and index.
Identifiers: LCCN 2022028978 (print) | LCCN 2022028979 (ebook) |
ISBN 9781138387201 (hbk) | ISBN 9781032375700 (pbk) |
ISBN 9780429426414 (ebk)
Subjects: LCSH: Pesticides—Application—New Zealand. | Pesticides—
Environmental aspects—New Zealand. | Moths—Control—
New Zealand. | Environmental health. | Urban health. |
Environmental sociology.
Classification: LCC TD196.P38 V35 2023 (print) |
LCC TD196.P38 (ebook) | DDC 363.738/4980993—dc23/eng/20221011
LC record available at https://lccn.loc.gov/2022028978
LC ebook record available at https://lccn.loc.gov/2022028979

ISBN: 978-1-138-38720-1 (hbk)
ISBN: 978-1-032-37570-0 (pbk)
ISBN: 978-0-429-42641-4 (ebk)

DOI: 10.4324/9780429426414

The publication of the Open Access version of this book was funded by
University of Auckland

Contents

Acknowledgements

A project of this nature is always enabled by scores of people who provide inspiration, support, insight, tools, dispositions and knowledge, and whose contributions would remain invisible but for opportunities like this to acknowledge them.

First and foremost, I thank my mom (Louise Vallée-Rai) and dad (Krishan Rai) for instilling in me a deep appreciation of education and for always ensuring we had the financial means to pursue it. To my dad I owe additional thanks for passing on the value of hard work and perseverance, and to my mom I owe additional thanks for keeping us healthy, well-fed, as well as opening my eyes to social injustices and emphasizing the importance of questioning authority.

Regarding the questioning of authority, I am also grateful to Diane Clement, Joseph Ball, Anne-Sita Vallée-Rai, and Alise Cappel, each of whom, through lengthy discussions about pesticides and toxicants, have contributed to my thinking on the subject. I also thank Professor Laura Nader for helping me (through her profound *Controlling Processes* course) draw back the curtain on the manifold social forces that have pushed, prodded, and manipulated me and my fellow citizens throughout my life, and for exposing the strategies and processes through which they sought to accomplish their objective.

I owe Tracey McIntosh, David Craig, and Vivienne Elizabeth a big thanks for giving me the opportunity to come live and work in beautiful Aotearoa. Besides being a wonderful opportunity to discover a new land, people, and culture, coming to Auckland gave me the opportunity to discover New Zealand's spraying operation against the Painted Apple Moth, which is the focus of this book. I also thank my departmental colleagues for all they have done over the years to maintain a positive and supportive workplace. I, for one have appreciated it very much.

A huge thank you goes out to Brittany Whiley and Florence Reynolds, who were the Summer Scholars appointed early on to my project. Both of them were a delight to work with, and both were fantastic at finding key documents and helping to formulate preliminary answers to key questions. Their scholarship was first rate and their efforts created a strong foundation for me to build on. As well, I thank the Faculty of Arts, at the University of Auckland, for funding their salaries. Thanks also go out to the Faculty and University leadership, for having the wisdom to continue supporting the Summer Scholars program, despite the financial pressures we have faced over the last several years. The program has been a boon to early career academics and aspiring graduate students, and I hope the program will continue for years to come.

A crucial component of such projects are the people who have offered access to key resources along the way. I offer a special thank you to my numerous informants, who were generous with their time and energy and whose contributions helped deepen my understanding of the Painted Apple Moth (PAM) spraying operation. Thanks also go out to my university colleague Marie McEntee, who, early in the process, lent me her collection of PAM media articles and advertisements, which proved to be very helpful. Thanks Marie!

My heartfelt thanks go out to my writing buddies Tracey Perkins and Lauren Richter, who provided feedback on early drafts of several chapters, as well as much-needed camaraderie and moral support during our many Zoom meetings. Also, a big thank you goes out to the anonymous reviewers of my book proposal, who provided generous and insightful feedback.

Jill Harrison's work on pesticides proved to be quite inspiring, and I thank her for, early in the project, encouraging me to focus on the book. I also thank Michael Carolan for supporting the project early on, as well as helping me see the implications of this project for spraying operations against mosquitos and other potential disease vectors. Thanks also go out to Kevin Dew, who provided insightful feedback on my conference presentation about the PAM spraying operation and who also encouraged me to pursue the project. A special thank you goes out to my good friend and departmental colleague Steve Matthewman, who gave me a sense of what it was like to live through the pesticide sprayings and who always provided steady support for the project, as well as a continuous flow of much-needed humor.

Some of the ideas in Chapter 6 were previously published by Routledge in "Nurturing an acquiescence to toxicity: The state's naturework in urban aerial pesticide spraying campaigns," *Resilience, Environmental Justice and the City* (2016). Similarly, ideas from Chapter 8 were published by Taylor & Francis in "Fostering ignorance to maintain public support: New Zealand's 2002–2004 urban aerial pesticide spraying operation over Auckland," *Environmental Sociology* 4 (2021): 382–292. I thank the anonymous reviewers of both articles for their generous and helpful feedback, which helped strengthen my ideas on each set of issues. As well, I thank the editorial staff at Routledge Press (in particular, Matthew Shobbrook, Rosie Anderson, Leila Walker, and Rebecca Brennan), who have been very supportive throughout the process and have been delightful to work with.

Finally, I'm deeply grateful for my family. I thank my supportive and ingenious wife (Alise Cappel) for not only being open to the idea of moving to New Zealand but for actively encouraging us to pursue this opportunity. As well, I thank her for providing eagle-eyed proofreading on several chapters, for always being there to bounce ideas off of, for providing moral support, for keeping me well-nourished with healthy and delicious meals, and for helping me laugh on a regular basis, which is more important than we realize. As for my son, he is a delight to have in my life and his presence inspires me to both be hopeful about the future and envision ways to build societies that are just, resilient, and sustainable.

Introduction

This book is about the social processes that lead to pesticide use. Pesticides are substances used to repel, kill, and control plants and animal life that are considered pests at a given place and time. These substances include insecticides for killing a range of insects, herbicides for eliminating weeds and other unwanted plants, fungicides for preventing mildew and mold, and rodenticides to control rats and other rodents. The use of chemical pesticides, in particular, has grown sharply in the post-WWII decades. In the United States alone pesticide use grew from 196 million pounds of pesticides in 1960 to over 1.1 billion pounds in 2012 (Atwood & Paisley-Jones 2017; Fernandez-Cornejo et al. 2014). As well, by 2012 worldwide pesticide use had surpassed 6 billion pounds (Atwood & Paisley-Jones 2017).

While much of the increase has occurred in sparsely populated rural areas, there has also been an insidious growth of pesticide use in densely populated urban environments, including parks, golf courses, home gardens, and even homes, in order to ward off unwanted insects and plants. Additionally, when faced with invasive species, governments have resorted to large-scale aerial pesticide spraying operations in urban areas, a phenomenon that will be the focus of this book and which I return to below.

There are several reasons to be concerned about the expanding use of pesticides. First, they have polluted wide swathes of human life, as pesticides increasingly contaminate the spaces we inhabit, the air we breathe, the water we drink, and the food we eat. As a result of such chemical trespasses, humans bioaccumulate chemicals in their organs and other tissues, with biomonitoring studies consistently finding evidence of contamination in most citizens. For instance, in 2005 the U.S. Centers for Disease Control reported that 90% of U.S. residents carry a mixture of pesticides in their bodies and that chlorpyrifos, in particular, was found in 75% of people who were tested (PANNA 2005).

In turn, such bioaccumulation predisposes people for long-term disease. On this point, researchers have found that pesticide exposure is linked to numerous medical conditions, including: (1) neurological problems, such as depression; (2) Parkinson's disease; (3) male fertility problems, such as lowered sperm counts; (4) female infertility; (5) birth defects; (6) endocrine

DOI: 10.4324/9780429426414-1

disruption; (7) respiratory problems, such as asthma; and (8) and cancers, including breast, prostate, and lung cancers (Bell 2012; CHE 2016a; Colburn et al. 1996; Pimental & Burgess 2014; Steingraber 2009). Researchers estimate that, in the United States alone, the human health impacts of pesticides surpasses 1.2 billion dollars per year (Pimental & Burgess 2014).

While these issues are particularly concerning for farm workers, who are routinely exposed to pesticides, they are also concerning for those living near farms due to frequent pesticide drifts (Harrison 2011). Concerns also extend to food consumers, as food produce is usually tainted with pesticide residues. For instance, one study found that 73% of fruits and vegetables had pesticide residues (Baker et al. 2002), with the rate surpassing 90% for apples, pears, peaches, strawberries, and celery. The issue is particularly concerning for children because (A) they consume more food and pesticides per weight than adults; (B) their detoxification pathways are often undeveloped; and (C) their brains are more than five times larger, in proportion to body weight, than adult brains, thereby making them more sensitive to the brain-harming aspects of pesticides (Pimental & Burgess 2014). Beyond endangering all who are exposed, pesticide-induced health risks are most often borne by the poor and people of color, who are the ones most often placed in harm's way (Bell 2012; Harrison 2011).

Beyond harming human health, pesticides also harm eco-systems in numerous ways. As Rachel Carson (1965) originally pointed out, chemical pesticides act indiscriminately, killing beneficial insects as well as the target species. Some beneficial insects are natural predators of the target insects, and eliminating them weakens ecosystem resilience and increases ecosystem vulnerability to pest infestations, which, in turn, increases the farmers' reliance on pesticides. Pollinators (such as bees and butterflies) are also harmed by pesticides, which hinders their ability to provide ecosystem services we rely on. Pimental and Burgess (2014) estimate that the pesticide harm to these insects leads to economic losses surpassing $283 million each year in the United States. Pesticides also harm soil-dwelling arthropods, fungi, bacteria, earthworms, and protozoa, each of which also performs important ecosystem services.

Beyond killing insects, pesticides make their way up the food chain, becoming more concentrated in the higher order species. As Carson (1965) illuminated, birds have been particularly imperilled by this process, with insect and worm-eating birds dying in significant numbers in areas where pesticides had been applied. This was particularly true of predatory birds (such as bald eagles, kestrels, peregrine falcons, and ospreys) and other birds at the higher end of the food chain (Lincer 1975; Pimental & Burgess 2014). While the problem was somewhat addressed by the banning of DDT in 1972, the problem has persisted. For instance, it is estimated that in 1996 the application of pesticides in Argentina farm fields led to the deaths of over 20,000 hawks. Moreover, it is estimated that the use of carbofuran kills 1 to 2 million birds each year in the United States (Pimental & Burgess 2014). As well, the

American Bird Conservancy (2010) estimates 12 particularly harmful pesticides (including fenthion, ethyl parathion, and chlorfenapyr) are responsible for killing over 15 million birds each year in the United States. Diminishing bird populations further weakens ecosystem diversity and resilience, which also makes farmers both more vulnerable to pest infestations and more reliant on chemical pesticides to ward off such infestations.

Pesticide use also has other significant eco-system effects, including poisoning the soil and making it unsuitable for planting other crops, soil erosion, pollution of ground and surface water, fish kills, harming the reproductive potential of birds and other mammals, and increased pesticide resistance in pests (Hallberg 1987; Pimental & Burgess 2014; Reganold et al. 1987). In turn, each of these further undermine the sustainability and resilience of human communities.

Social Science Research on Pesticide Use

Although Rachel Carson's *Silent Spring* (1965) revolutionized people's understanding of pesticides and helped spark a movement to reduce specific hazards, the use of pesticides and industrial chemicals has grown tremendously, leading some scholars to argue chemical toxicity now reaches more deeply and broadly than it ever did in Carson's era (Woodhouse & Howard 2009).

This book's overarching aim is to elucidate social processes that contribute to pesticide use. Previous research in this area has identified numerous contributing factors, with some focusing particularly on the role of farmers (Guivant 2003; Lockie 1997; Ward 1995). For instance, Stewart Lockie (1997) examined the Australian farmers' rational for continuing to use pesticides and found that they view pesticides as a hedge against short-term risk in the volatile commodity markets, which neoliberal governments are increasingly exposing them to. Additionally, he found that farmers view weed control herbicides as a means of preserving their status as environmentally and socially responsible actors, due to the perception that such herbicides are less impactful than mechanical cultivation. Elsewhere, Neil Ward (1995) examined the expansion of herbicide use in the United Kingdom and elucidated how farmer decisions were informed by industry technical advisors and government institutions that defined "good farming" practices as the elimination of all weeds. Ward's (1995) analysis emphasizes that increasing pesticide use is not due to technological rationalism but is rather socially constructed by proponents (such as manufacturers) who work to convince potential users to deploy pesticides and labor to shape institutions (such as safety regulations) to accord with their agenda.

Another research stream examines the way pesticide use is linked to institutions, with weak regulations being a particularly important one. Regulatory failures are caused by various factors, including: policy stasis that has prevented meaningful progress on the issue (Bosso 1987); regulatory capture (Daniel 2005; Harrison 2006, 2011); the devolution of environmental governance to the local level (Harrison 2006); the ability of regulatory officials to portray pesticide problems as local problems requiring local solutions, instead of requiring state- or federal-level interventions (Harrison 2006); regulator reliance on inappropriate assessment tools (Wargo 1996); and the "epistemic forms" (i.e. the professional rules, procedures, and norms that shape knowledge construction in a given field) that lead regulators to use assessment tools that are ill-equipped to fully capture the damage pesticides do to ecosystems (Kleinman & Suryanarayanan 2013).

At a more macro level, Jorgenson and Kykendall's (2008) work highlights the political economy's contribution, by elucidating how pesticide use in developing countries is mediated by their level of dependence on foreign investment. Culture is another structural factor, and Harrison (2014) explains how California's high pesticide use can be traced to mainstream environmentalism's use of libertarian and communitarian ideas of justice, whose focus on consumer benefits obscures concerns for farmworker well-being. Combining a focus on culture and political economy, Shorette's (2012) work reveals how pesticide use is mediated by a country's degree of integration with world cultural institutions and its place in the larger world system.

Another research stream has studied the public's acceptance of pesticide use, which has consistently found that public support for pesticides is highly correlated with perceptions that pesticides are safe (Coppin et al. 2002; Dunlap & Beus 1992; Johnson et al. 1995). Pushing the analysis further, Coppin et al. (2002) found that safety perceptions were correlated with educational attainment, concerns about pesticide exposure, trust in the pesticide industry, and trust in the information provided by federal agencies.

Beyond illuminating individual-level determinants of public attitudes, research has also illuminated how social forces (i.e. industry, media, and government) have shaped public attitudes about pesticides. For example, Michelle Mart (2015) illuminates how 1950s U.S. mainstream media consistently portrayed pesticides in a laudatory and uncritical light, thereby encouraging the public to admire the "wonders" of modern chemistry. Governments can also work to shape public perceptions, as emphasized by Harrison's work (2011), which illuminates the tendency of California government officials to routinely dismiss the human impacts of pesticide drift. While Harrison's (2011) contribution is a good start, more attention needs to be directed to studying government's role. In particular, researchers need to identify the other strategies and tactics that government agents use to allay citizen concerns about pesticide safety.

Another important issue is the process through which pesticide proponents build public support for pesticide use by framing targeted species as a social problem. This is a crucial issue to pesticide use because defining something as a social problem (such as being defined as weeds, pests, or perhaps even biosecurity threats) lays the groundwork for deploying pesticides to eliminate that problem. Thus, in order to develop a deeper understanding of pesticide use, it is imperative we elucidate the ideological work pursued to problematize certain species. A pertinent work on the issue is Ward's (1995) analysis of herbicide use in the United Kingdom, which illuminated the way herbicide use was tied to perceptions that "good farming practice" meant eliminating weeds. While useful, Ward's work did not document the ideological work through which weeds were problematized in the first place. Particularly pertinent in this regard is Michelle Mart's (2015) analysis of America's love affair with pesticides, which illuminates the extent to which industry and mainstream media portrayed insects as humanity's mortal enemy, which is a significant distortion of the important role insects play in our ecosystems. In a similar vein, Wildblood-Crawford's (2006) analysis of a New Zealand chemical industry trade journal reveals how it too has depicted insects as humanity's steadfast enemy. While such analyses help shed light on an important topic, an area they do not address is government efforts to portray insects in a negative light. This is a significant gap as government agencies have significant resources to develop negative portrayals of insects, have privileged access to the engines of mass media, and are often perceived as an authoritative source of information. This means they can significantly increase the likelihood that citizens will view an insect as a "pest" needing to be eradicated, which, again, lays the groundwork for deploying pesticides.

Urban Biosecurity Pesticide Spraying

To shed deeper light on these issues, this book examines the issue of aerial pesticide spraying campaigns carried out over urban areas in order to eradicate foreign species. While most people are familiar with agricultural pesticide spraying, far fewer are familiar with the 20 and counting pesticide spraying operations that Canadian, American, and New Zealand governments have pursued over urban neighborhoods since 1993, including Spokane, WA (1993); Auckland, New Zealand (1996–1997 and 2002–2004); Victoria, British Columbia (1998); Seattle, WA (2000 and 2016); Hamilton, New Zealand (2003–2004); Charlotte, North Carolina (1992, 1998 and 2008); San Francisco Bay Area, CA (2008–2010); Toronto, Ontario (2007, 2008, 2013, 2017, 2019, and 2020); and Surrey, British Columbia (2015, 2019, and 2020) (Associated Press 2016; DH Toronto Staff 2019, 2020; Johal 2020; Office of the Ombudsman 2007; Sumlin 2015; Washington State Department of Health 2001; Zargar 2019).

Although these operations are geared towards eradicating biosecurity threats, they raise important public health and ecological considerations as the pesticides they use can cause collateral damage to humans, other species, and ecosystems. For example, the Foray 48B pesticide, which has been used in many spraying operations, has synthetic chemicals that can cause human health problems when inhaled, including bronchitis, asthma, lung cancer, intestinal damage, depression, and inflammation of the respiratory tract (Brunekeef & Holgate 2002; CHE 2016b; Patnaik 2007; Vizcaya et al. 2011).

As for environmental impacts, there is evidence to suggest the sprays can exact a high toll on ecological diversity. For instance, the main pesticide ingredient used in the Auckland spraying operation (i.e. the Btk bacteria) acts indiscriminately on all caterpillars, and so its use can cause collapses of native caterpillars. Indeed, a 1994 Btk spraying in Oregon was found to have reduced by 80% the number of local caterpillars and reduced by 60% the number of caterpillar species (Savonen 1994). In turn, this reduced the ecosystem services provided by such species, including pollination, weed control, and serving as a food source for birds and other vital elements of the ecosystem (Swadener 1994).

There are several compelling reasons to study urban biosecurity spraying operations. First, it is an understudied social phenomenon. Although social scientists have written extensively about agricultural pesticide use (Bosso 1987; Harrison 2011; Kleinman & Suryanarayanan 2013; Wargo 1996), little has been written about urban spraying campaigns. Second, the numerous people who are exposed and the indiscriminate manner of exposure raise significant public health concerns that need to be better understood. As well, indiscriminate pesticide spraying from airplanes raises important environmental concerns.

Additionally, studying these spraying operations provides a strategic opportunity for revealing the social processes through which governments nurture public acceptance of pesticide spraying. Garnering public acceptance is crucial for all pesticides as its absence impedes industry attempts to drive usage. Not only will consumers be more reluctant to use pesticides, but also they are likely to pressure for tougher regulations and bans, as occurred in Québec (Canada), where citizen pressure led Montréal city officials to ban glyphosate at the end of 2019 (Montreal Gazette 2019) and provincial officials to issue a province-wide ban of atrazine, chlorpyrifos, and three neonicotinoids (clothianidin, imidacloprid, and thiamethoxam) in 2018 (Fletcher 2018). However, while public acceptance is important to all pesticide use, it is particularly important to the biosecurity urban spraying operations, as exposing large numbers of people to pesticides can trigger significant political backlash.

Beyond helping us better understand factors that contribute to the normalization of pesticide use, studying biosecurity spraying operations can help us better understand key issues about biosecurity, including (1) the social factors that contribute to the incursion of invasive species; (2) the social process through which an invasive species comes to be seen as a biosecurity concern or threat; (3) the role government agencies play in that process; (4) who benefits the most from the framing process; and (5) how these issues are mediated by the capitalist political economy.

The Case: New Zealand's 2002–2004 Painted Apple Moth Eradication Operation

The specific case analyzed in this book is the aerial pesticide spraying operation that New Zealand government officials conducted over Auckland, from January 2002 to May 2004, in order to eliminate the painted apple moth. Many features make this case particularly strategic to study. First, it was the most extensive urban spraying campaign ever undertaken in the world, whether considering its duration (29 months), the number of sprayings administered (more than 45), the size of the spray zone (10,632 hectares (26,272 acres) at its peak, which equates to 14,890 football fields), or the number of people exposed (193,000 to 300,000) (Goven et al. 2007; Office of the Ombudsman 2007). Second, surveys reveal the campaign enjoyed solid public support at its outset, with 70% initially supporting aerial pesticide spraying if necessary (Office of the Ombudsman 2007). Such broad support is intriguing. It is one thing for urbanites to accept pesticide spraying in sparsely populated rural regions, which are out-of-sight and out-of-mind but quite another to accept aerial pesticide spraying in their own backyards, where it is viscerally experienced – where people see it being sprayed, smell it in gardens and inside homes, feel it on skin and in lungs, and hear helicopters and planes approach for the next spraying.

Third, despite the unprecedented number of both pesticide sprayings and number of people who were repeatedly exposed, opposition to the spraying spread quite slowly. While some public opposition emerged at the outset, it was quite slow to spread, as underscored by the fact that 67% of residents still supported the operation ten months into it and that only 13% did not (Office of the Ombudsman 2007). As a result, the government was able to continue its spraying operation until its end, which was 24 months longer than the 5 months operation it proposed at the outset. It is possible the public's initial support was due to their lack of experience with spraying operations and an inability to fully comprehend what they were endorsing. However, it is another matter to continue supporting pesticide spraying one is viscerally experiencing, month after month. The slow spread of opposition is even more intriguing when we consider the large numbers who were repeatedly exposed and that nearly 2% of those (3,800 residents) were affected enough to seek medical assistance (Office of the Ombudsman 2007). These factors, coupled with the fact urban-dwellers are more likely to possess the political, economic, and cultural capital needed to effectively oppose pesticide campaigns (Harrison 2011), should have set the stage for a large-scale social backlash. However, such a backlash was quite slow to materialize, which suggests the public's response was managed by powerful social forces.

This case raises numerous questions that will get addressed in the book, including the following: (1) how did social networks, social systems, and human activities contribute to PAM's arrival and spread in New Zealand?; (2) why was New Zealand so intolerant of this foreign species?; (3) how did the government build public support for eradicating the moth?; (4) why wasn't there more public opposition to the spraying operation?; (5) how did government officials allay public concerns about the spraying?; (6) how was the public's response mediated by contextual factors?; and (7) what accounts for the government's willingness to both expose hundreds of thousands of people to a pesticide spray and mislead the public about the health and environmental concerns associated with it.

The Conceptual Framework

In addressing these questions I employed a synthetic conceptual approach, which drew on contributions from several key literatures.

Regarding the first question, I drew on the powerful environmental sociology insight that environmental problems do not simply occur but rather are significantly mediated by human activities and social systems. For instance, in their analysis of Hurricane Katrina, McCarthy and King (2014) emphasize that "while hurricanes are typically considered 'natural disasters,' Katrina's extreme consequences must be considered the result of social and political failures"(p. 4), including the failure to ensure satisfactory storm protection plans, the US Army corps of engineers' use of outdated data in building levees and floodwalls, and the reduction of wetlands, which normally serve as a natural buffer against hurricane storm surge. Similarly, they argue that the United States' air pollution problem has roots in many social processes, including the "inability of governmental policies and laws to regulate industry amidst the rise of neoliberal style economics, with its emphasis on deregulation and corporate rights, and the increasing monetary and political power of corporations" (p. 5). In a similar vein, geographers emphasize that the migration of species is mediated by socio-biological networks, which includes networks of exchange and social systems interacting with those networks, and that our focus should be on illuminating the "human ecologies of species invasions," instead of fixating solely on the species invasions that result from those ecologies (Jay & Morad 2006; Robbins 2004). The "social production of environmental problems" perspective suggests all environmental problems should be situated within the web of human actions and social systems that created them and/or mediated their intensity. Correspondingly, this is the framework I used to make sense of PAM's arrival in New Zealand.

In addressing the second question I utilized a social constructionist approach, which argues that the recognition of an environmental problem is the end-product of a "dynamic social process of definition, negotiation and legitimation" (Hannigan 1995, p. 31). An example of a constructionist analysis is Hannigan's (2014) discussion of biodiversity loss, whereby scientists assembled the claim and then persuasively presented it to the larger public. In a similar way, the constructionist lens can help us make sense of "biosecurity crises," such as the PAM incursion, where proponents of the "crisis" frame have to assemble their claim and persuasively present it to the larger public. Another useful tool is Gary Alan Fine's (1988) "naturework" concept, which refers to the ideological work involved with turning nature into culture. That is to say, it is the ideological work that shapes the cultural lenses through which we interpret nature and which determines how a natural entity is viewed. In this case, I paid close attention to "naturework" government officials undertook to frame PAM as a biosecurity threat. Beyond elucidating the social processes through which the PAM incursion came to be seen as a biosecurity "crisis," I contextualize the case by tracing the economic, cultural, and historical setting within which it unfolded.

The constructionist lens is also useful for understanding why pesticides were chosen as the solution to the PAM incursion. A constructionist lens encourages us to see the chosen solutions as the end-point in a complex social process, which is also mediated by cultural history, all of which needs to be unpacked and analyzed. While part of that analysis consists of identifying why a particular solution was chosen, another part consists of analyzing why opponents to the solution failed to build up sufficient opposition. As it relates to the PAM case, a constructionist approach encourages us to examine why opposition to the spraying was so slow to build.

In trying to account for the slow spread of opposition, I drew on Laura Nader's (1997) "controlling processes" concept, which she defines as micro-processes through which certain conditions become normalized and through which "individuals and groups are influenced and persuaded to participate in their own domination" (p. 712). She argues that such processes are particularly prevalent in industrialized democratic societies, where coercive power is less culturally acceptable and where power is increasingly exerted through cultural controls, which channel taste, values, and behavior. Nader's work encourages us to identify the social forces operating in a given context and to analyze the way their actions curtail potential resistance. As it pertains to urban pesticide spraying operations, government agencies are a significant social force, as they either administer the spraying themselves or employ firms to do it. As will get revealed over the course of this book, there were several controlling processes that helped reduce the likelihood citizens would oppose the PAM spraying operation, which included the use of fear-mongering and the "biosecurity" discourse.

Attenuating Risk Perceptions

The previous literature has made it quite clear that a strong mediator of public support for pesticide use is the safety perceptions surrounding those products (Coppin et al. 2002; Dunlap & Beus 1992; Johnson et al. 1995). To build on those insights, I drew on the branch of risk studies that examines the amplification and attenuation of risk perceptions of potentially harmful products. While much of this literature has focused on the way activist groups and media try to amplify risk perceptions around particular products (Henderson et al. 2014; Kasperson & Kasperson 1996), Marc-Olivier Déplaude (2015) takes a different tack, analyzing the social processes through which *industries* seek to attenuate risk perceptions about their products. His particular focus is on the French salt manufacturers, who used various tactics (dissimulation, denial, diversion, undermining opponents, and intimidating opponents) to reduce the risk perceptions associated with salt consumption. In illuminating this case, Déplaude (2015) encourages us to pursue similar analyses for other potentially harmful substances, such as pesticides, while also providing a useful preliminary framework for carrying out such analyses. Correspondingly, I analyzed the processes through which government officials attenuated risk perceptions regarding the PAM pesticide, which included proposing a much smaller operation than would eventually get carried out and expanding the operation incrementally.

Another means of attenuating risk perceptions is through the production of ignorance, and the burgeoning literature on this topic supplied numerous insights. First, it emphasizes that while ignorance can consist of the absence of knowledge about a topic, it can also consist of false knowledge, where people hold erroneous information about a topic and/or give disproportionate attention to marginal or industry-funded research, as has been the case with climate change deniers (Michaels 2008; Proctor 2008). Second, it emphasizes that ignorance is more than a knowledge gap to be filled or a set of incorrect ideas to be corrected, as ignorance can also be a resource for those in power, cultivated to serve strategic purposes (McGoey 2012; Oreskes & Conway 2010; Proctor 2008; Rayner 2012). For this reason, McGoey (2012) argues social scientists should focus less on the politics of knowledge and more on the

> politics of ignorance, the mobilization of ambiguity, the denial of unsettling facts, the realization that knowing the least amount possible is often the most indispensable tool for managing risks and exonerating oneself from blame in the aftermath of catastrophic events.
>
> (p. 3)

In turn, the deliberate cultivation of ignorance signals that some knowledge is inconvenient or unsettling, what Rayner (2012) refers to as "uncomfortable knowledge." The literature also illuminates how such knowledge can be obscured from the public view, through failing to carry out necessary science (i.e. undone science) and/or deploying neutralization tactics, such as suppression, omission, dismissal, or diversion.

Undone science, as illuminated through the work of Frickel et al. (2010) and Hess (2007), refers to the failure to authorize, fund, and/or complete research that some stakeholders consider to be essential. Undone science is particularly important for industries as the absence of knowledge about harms removes an obstacle to portraying their products as safe, which enables them to more effectively resist regulations. An example is the pharmaceutical industry's failure to pursue long-term safety testing of Ritalin, a controversial psychostimulant given to children for attention deficit hyperactivity disorder. The deliberate refusal to conduct long-term safety studies perpetuates non-knowledge about the medication's long-term safety, which continues to obscure and leave unanswered key questions about its safety. Conveniently and ironically, proponents of Ritalin and competitor drugs have used appeals to ignorance (appeals to the non-existence of this knowledge), as support for claims that Ritalin is safe, and to fend off additional regulations. Appeals to ignorance, however, are logically deficient. The absence of evidence is not evidence of absence. Absence of evidence can be caused by several factors, including the failure to look for evidence of the problem or failure to do so effectively. In turn, these problems open the door to declaring false negatives (i.e. proclaiming there is no problem when the opposite might be the case). This dynamic is particularly germane to the PAM case, as two of the government's central tactics were to portray the pesticide as harmless and to base those claims on appeals to ignorance, as I detail in subsequent chapters.

When uncomfortable knowledge does get produced, suppression becomes a key tactic to conceal it from the public. One suppression tactic is intimidation, as Déplaude (2015) explores in his analysis of the French salt industry, which threatened legal action against scientists who produced inconvenient knowledge about salt's harmfulness (Déplaude 2015). As Galison (2008) emphasizes, other powerful suppression tactics include censorship and designating something as "classified" knowledge. The latter is particularly pertinent to the PAM case, as the government refused to reveal the synthetic chemicals in the PAM pesticide, thereby preventing the public from knowing what chemicals they were being exposed to or how to properly protect themselves against those chemicals.

When suppression does not work, those in power tend to pivot to other neutralization tactics, such as disseminating information that omits inconvenient knowledge. Researchers have documented the use of omission by a range of different industries, including tobacco, lead, asbestos, chemical, and pharmaceutical (Healy 2012; Markowitz & Rosner 2002; McCulloch & Tweedale 2008; Moyers et al. 2002; Proctor 2008; Vogel 2012). While the strategy seems to be de rigueur among industry actors, it can also be used among government agencies to advance their objectives, as I discuss in later chapters.

Another neutralization strategy is dismissal, which Rayner (2012) defines as engaging with uncomfortable knowledge, in order to rebut it. A particularly potent industry rebuttal tactic is to create and sow doubt regarding the injury claims against their products (Proctor 2008). For instance, over the decades, tobacco manufacturers have downplayed the significance of uncomfortable knowledge, by arguing that evidence is far from conclusive and needs to be supplemented with more research (Proctor 2008). A related dismissal tactic is disseminating public statements that disproportionately highlight or exaggerate uncertainty in opposition research, as exemplified by the French salt industry's antics (Déplaude 2015). Another dismissal tactic is portraying findings as only relevant for a small subset of people, a tactic pharmaceutical manufacturers routinely employ in their US television ads. Yet another dismissal tactic is suggesting the problem represents insignificant risk to those affected by it. Dismissal is another strategy the New Zealand government used extensively.

Diversion refers to actions pursued to draw the public's attention away from the product's potential harmfulness. Proctor (2008) illuminates the great lengths tobacco manufacturers took to divert the public's gaze from the relationship between tobacco and cancer. This included funding research to divert attention to other potential causes of lung cancer and hiring historians to create a positive narrative of the tobacco industry. While the New Zealand government did not use these tactics, they used other diversionary tactics, including associating the pesticide with nature and organic farming.

Beyond identifying how government sought to attenuate risk perceptions, it is also important to consider how the efficacy of those activities was mediated by the larger cultural context. To shed light on this issue I drew on Edward Woodhouse and Jeff Howard's (2009) work on the underlying ideological structures that enable the growing use of toxic products in first world countries. Their particular focus is on the role played by "governing mentalities," which they define as "a tacit and often ill-considered pattern of assumptions that fundamentally shapes political relationships, interactions, and dialogue, often in ways that conflict with democratic ideals" (*Ibid.* p. 46). They argue that society's rampant use of harmful toxicants is enabled by three governing mentalities: (1) granting business executives the authority to decide which products to produce; (2) granting academics the authority to decide what to research and teach regarding technologies; and (3) the citizens' fatalistic acquiescence to toxicity, where the acceptance of toxicants is seen as the necessary price to pay for living in an affluent consumer society. As I will show in subsequent chapters, these governing mentalities were quite pertinent to the PAM case. Another mediating contextual factor is the education that is provided, as being properly educated about toxicants, mass communications, and the state would make citizens more able to defend themselves against government manipulation. Conversely, educational deficiencies on these key issues make it easier for governments to manipulate and control citizens, as was the case in the PAM case.

Conflict Theories of the State

Beyond illuminating the activities that the government carried out during the spraying operation, this book tackles the important question of why government agencies would expose hundreds of thousands of citizens to potentially harmful substances and then mislead those citizens about the concerns surrounding those substances. For guidance on this issue I drew on the work of conflict theorists, who provided several important insights. First, conflict theorists eschew pluralist conceptions of the state, arguing that the state is not a neutral arbiter of social conflict but rather is a resource controlled by dominant groups (Buechler 2014). In a capitalist political economy industry executives represent the dominant group and we should expect the state to pursue actions and policies that will benefit industry, even if it comes at the expense of environmental and public health.

While some neo-Marxists have an instrumental theory of the state, others offer a structuralist model, where the state does not pursue industry-protecting actions because it has been captured by capitalists, but rather because the state is oriented towards identifying and preserving the general and long-term interests of capitalism (Buechler 2014). Synonymous with this perspective is the work of Schnaiberg and Gould (1994), whose "treadmill of production" concept suggests that governments pursue ecological and public-health harming activities because they are addicted to economic growth and the tax revenue it brings in. In this perspective, the problems can be traced to ideology: the system of beliefs and values that shape how politicians interpret the world, their role within it, and the type of solutions they are predisposed to pursue.

Also useful is Weber's definition of the state as an entity that "successfully claims a monopoly on the legitimate use of physical force" (Buechler 2014, p. 74). Weber's definition can be usefully extended to include the imposition of urban aerial spraying operations, which can be seen as another manifestation of state violence. Also useful from his definition is its emphasis on legitimacy, which underscores the important role the public plays in enabling state violence. As Buechler (2014) argues, "legitimation leads people to accept and even support state violence while rejecting violence done by others" (p. 74). Flowing on from these insights, while I sought to better understand the ideologies shaping government officials, I also sought to understand the reasons the public granted legitimacy and authority to those officials.

Book Structure

Chapter 1 is meant to give readers an overview of the PAM eradication operation. Towards this end, it identifies key events in the case, including: (1) the moth's initial detection in May 1999; (2) the government's unsuccessful initial attempt to eradicate PAM; (3) its subsequent decision to pursue an aerial spraying campaign; (4) the start of aerial spraying in January 2002; (5) the 2003 release of critical health reports by Meriel Watts and then Hana Blackmore, which put considerable pressure on the government to assess the spraying's human health impact; (6) the Ministry of Health's commissioning of researchers, in March 2003, to assess those health effects and the Ministry's efforts to delay the production and release of that knowledge; (7) the Ministry of Agriculture and Forestry's decision to re-extend the spraying operation beyond the extended May 2003 end date; and (8) the end of spraying in May 2004. Beyond mapping the key events in this story, the chapter introduces readers to some of the key players, while also identifying the environmental and health concerns associated with pursuing an extensive aerial pesticide spraying operation over urban neighborhoods.

Chapter 2 focuses on how foreign species incursions occur. Based on the principle that environmental issues are always mediated by human activities and social systems, the chapter situates the PAM incursion within a web of social systems and human activities, elucidating the social factors that contributed to creating and exacerbating the PAM incursion. An important enabling factor was the late twentieth-century rise in global trade, which increased the opportunities for species from one locale to migrate to new ones. Compounding this problem were technological developments that decreased travel times and increased the likelihood a foreign species would survive the trip. Another mediating factor was the New Zealand government's reluctance to adequately fund biosecurity border controls, which increased the likelihood that a stowaway species could establish itself in New Zealand. These problems were further compounded by the government's ineffective initial response to PAM, which allowed the moth population to grow and spread considerably.

Chapter 3 examines why government agencies decided to target PAM for eradication, which was not a foregone conclusion. Over the last 180 years, New Zealand has experienced the arrival of over 19,000 new species and not all have been targeted for eradication. In fact, the importation of many invasive species (like rabbits, deer, ferrets, gorse, and pine trees) were enthusiastically encouraged at certain moments in history, and some (such as cats and dogs) continue to be tolerated, despite the immense damage they do to local fauna. The variation in response underscores that targeting a species for eradication is a complex social process that needs to be unpacked and situated in its cultural, historical, economic, and political context. This chapter examines the social forces that mediated PAM's reception and argues the government's response can be partially traced to the prevailing worldview, which conceptualizes the environment as a resource base to satisfy human needs and wants. The response can also be traced to the country's long history of fending off invasive species and the economic and ecological costs resulting from those incursions. While the impact of those incursions was often exacerbated by the lack of predators to keep invasive species in check, the impact was further heightened by agriculture and forestry's reliance on scientific management, which focuses on choosing the fastest growing species at the expense of building resilience against pests. In forestry's case, they have relied on the fast-growing pine trees for approximately 90% of its stock, which meant the industry is particularly vulnerable to any virus, fungus, or insect that would target those trees. While pine trees was not PAM's preferred food source, the forestry industry felt threatened by the insect because there was evidence suggesting it could feed on pine trees if its preferred food source was absent.

Chapter 4 examines why the government turned to an aerial spraying operation to eradicate PAM. It is not a given that a government agency will seek to control a species through a prolonged aerial pesticide spraying operation over urban neighborhoods, as other alternatives exist. For instance, vegetation controls can be implemented to prevent the spread of infested plant matter to other parts of a city, parasites or predators can be found to control the species biologically (it appears PAM was susceptible to viruses, parasitic fungus, and parasitic wasps (WASP 2002)), pheromone traps can be used, and mating disruption technologies (such as sterile moths) can be deployed (as was the case in Auckland's 1996–1997 eradication operation against the white tussock moth) (Walsh 2003). Thus, the government's choice to pursue aerial pesticide spraying needs to be analyzed and explained. The chapter contextualizes the pesticide response by situating it within the "synthetic age" (Foster 1999) we live in, where ecological problems tend to be addressed through technological interventions instead of ecological ones. This is particularly true in New Zealand, which has a long history of resolving ecological problems through pesticides, even when such approaches can harm human populations. The analysis also situates the pesticide response within the country's capitalist political economy, where there has been a long track record of prioritizing economic growth at the expense of human health. The chapter also considers how the stage was set by Auckland's 1996–1997 pesticide spraying operation to eliminate the white spotted tussock moth, which also took place in densely populated urban neighborhoods and set an important precedent. Lastly, the chapter examines MAF's bungled initial response to PAM, which allowed the moth to spread far and wide and alarmed forestry industry officials, who increasingly pressured MAF to administer aerial pesticide spraying over Auckland suburbs.

Governments do not operate in a social vacuum, and pesticide spraying operations can get derailed by citizen opposition. So, in trying to understand what enables pesticide spraying activities in urban areas, it is crucial to consider how citizens respond to proposed pesticide spraying operations. Towards that end, Chapter 5 traces Aucklanders' response to the aerial pesticide spraying operation. Part of that story includes a core of citizens who opposed the spraying through numerous means, which included advising government officials about community concerns and alternative options, publicly disseminating information that criticized the spraying operation and organizing protest marches and rallies. Despite those opposition activities, the community was unable to prevent the government from either carrying out the spraying operation it initially proposed or considerably expanding that operation over time. The chapter begins to illuminate the source of the opposition's ineffectiveness, which included a loophole in the environmental regulations that allowed government officials to essentially bypass the need for local consent. Another important factor was that while a dedicated core of locals did everything they could to oppose the spraying, that opposition did not spread quickly to the masses. The slow spread of opposition was a crucial element in the PAM case, and accounting for it is a significant focus of the remaining chapters. As part of that process, I drew on Laura Nader's (1997) "controlling processes" concept, by identifying processes through which government officials slowed the spread of opposition.

Chapter 6 reveals that one way they slowed the spread of opposition was to carry out fear-mongering ideological work that framed PAM as a triple biosecurity threat: i.e. a threat to native ecology, the economy, and public health. As well, it explains how the effectiveness of those efforts was enhanced by the surrounding cultural context, which included the country's long history of dealing with invasive species, the population's low eco-literacy, and the environment's importance to New Zealand national identity.

Besides building support for a pesticide spraying operation, proponents also have to work to allay citizen concerns about the pesticides in question. Chapter 7 examines several actions MAF pursued that helped allay citizen concerns about the PAM spraying operation, which included proposing a much smaller operation at the outset, expanding the operation by small increments, as well as consistently disseminating the narrative that the pesticide and its ingredients were harmless to humans.

Another important way to allay citizen concerns is to manage what Steve Rayner (2012) refers to as "uncomfortable knowledge," which is knowledge that contradicts the dominant narrative and which could undermine the government's political agenda. Chapter 8 details several strategies that government officials deployed. One was refusing to systematically assess the pesticide spraying's health impact on citizens. When emerging citizen science suggested people were being harmed by the spraying, government officials finally commissioned independent researchers to investigate the matter. However, the potential impact of that research was significantly reduced by the fact Ministry of Health officials significantly circumscribed the research project's scope. Moreover, they further limited its impact by delaying the production and release of the research findings. Another management strategy they used was neutralizing uncomfortable knowledge, either through suppression, omission, dismissal, denial, downplaying its significance, or diversion. Still another strategy consisted of neutralizing potential sources of uncomfortable knowledge, such as the Community Advisory Group, frontline medical staff, and the Ministry of Health.

Chapter 9 considers how the effectiveness of the government's communication campaign was mediated by the ideological context. Part of that context included the widespread acceptance of pesticide use, which inclined Aucklanders to view the PAM spraying operation as an acceptable means to eradicate the moth. The chapter also considers three factors that actively fed that cultural acceptance, which includes pesticide ignorance and what Woodhouse and Howard (2009) refer to as an "acquiescence to toxicity," which is the tendency of first world citizens to believe that exposure to toxicants is simply the price to pay for living in an affluent consumer society. The third factor feeding the acceptance is the education system's failure to adequately educate *all citizens* about pesticides and other toxicants. Besides the acceptance of pesticides, another key aspect of the ideological context is the university system's tendency to impart the deeply flawed pluralist conception of the state, which makes citizens much less likely to understand, let alone oppose, the way government agencies overwhelmingly favor corporate financial interests over population health and other citizen concerns. The chapter also considers the population's ignorance about the politics surrounding the production of knowledge and ignorance, which can also be traced to the education system and which makes citizens more susceptible to being manipulated by the type of tactics the New Zealand government deployed.

Lastly, the chapter illuminates the deeper ideological structures that give rise to the educational deficiencies, which include the widely accepted beliefs that universities should be organized as marketplaces of ideas and that the public should defer to the wisdom of academics for research and teaching taking place in the ivory tower.

The concluding chapter reflects on what New Zealand's PAM pesticide spraying operation can tell us about similar biosecurity spraying operations, as well as about other types of urban pesticide applications and pesticide use more generally. In particular, it highlights the state's proclivity for pursuing spraying operations that protect the maximization of industry profitability, even when those operations place human health and well-being at risk. State-sponsored chemical trespass is a real phenomenon, and the chapter calls for a more critical perspective on government agencies, one that views their actions as being intrinsically tied to the role they play in capitalist political economies. The chapter also discusses how citizens can better protect themselves against future unwarranted trespass from pesticides and other chemicals.

References

American Bird Conservancy. 2010. "Pesticides and birds." American Bird Conservancy. http://www.abcbirds.org/abcprograms/policy/toxins/pesticides.html

Associated Press. 2016. "State to spray over Tacoma, Seattle to kill gypsy moths." *KOMO News*, April 20. https://komonews.com/news/local/state-to-spray-over-tacoma-seattle-to-kill-gypsy-moths

Atwood, Donald and Claire Paisley-Jones. 2017. *Pesticides Industry Sales and Usage: 2008–2012 Market Estimates.* Washington, DC: US Environmental Protection Agency.

Baker, Brian, Charles Benbrook, Edward Groth III, and Karen Lutz Benbrook. 2002. "Pesticide Residues in Conventional, Integrated Pest Management (IPM)-Grown and Organic Foods: Insights from Three US Data Sets." *Food Additives and Contaminants* 19(5): 427–446.

Bell, Michael. 2012. *An Invitation to Environmental Sociology, 4th edition.* Thousand Oaks, CA: Pine Forge Press.

Bosso, Christopher. 1987. *Pesticides and Politics: The Life Cycle of a Public Issue.* Pittsburgh, PA: University of Pittsburgh Press.

Brunekeef, Bert and Stephen Holgate. 2002. "Air Pollution and Health." *Lancet* 360(9341): 1233–1242.

Buechler, Steven. 2014. *Critical Sociology, 2nd edition.* New York: Routledge Press.

Carson, Rachel. 1965. *Silent Spring.* Hammondsworth: Penguin Books.

CHE (Collaborative on Health and the Environment). 2016a. Toxicant and disease database entry for 'pesticides'. Accessed April 3, 2016. https://www.healthandenvironment.org/our-work/toxicant-and-disease-database/?showcategory=&showdisease=&showcontaminant=2380&showcas=&showkeyword=

CHE (Collaborative on Health and the Environment). 2016b. Toxicant and Disease database entry for 'propylene glycol'. Accessed April 3, 2016. http://www.healthandenvironment.org/tddb/contam/?itemid=2995

Colburn, Theo, Dianne Dumanoski, and John Peterson Myers. 1996. *Our Stolen Future: How We Are threatening our Fertility, Intelligence and Survival: A Scientific Detective Story.* New York: Dutton Press.

Coppin, Dawn, Brian Eisenhauer, and Richard Krannich. 2002. "Is Pesticide Use Socially Acceptable? A Comparison between Urban and Rural Settings." *Social Science Quarterly* 83(1): 379–393.

Daniel, Pete. 2005. *Toxic Drift: Pesticides and Health in the Post-World War II South.* Baton Rouge and Washington, DC: Louisiana State University Press in cooperation with the Smithsonian National Museum of American History.

Déplaude, Marc-Olivier. 2015. "Minimising Dietary Risk: The French Association of Salt Producers and the Manufacturing of Ignorance." *Health Risk & Society* 17(2): 168–183.

DH Toronto Staff. 2019. "There will be aerial spraying in parts of Toronto this weekend." *Daily Hive News*, May 23. https://dailyhive.com/toronto/aerial-spray-tree-damaging-european-gypsy-moth-toronto-may-2019

DH Toronto Staff. 2020. "There will be aerial spraying in parts of Toronto this week." *Daily Hive News*, May 26. https://dailyhive.com/toronto/aerial-spraying-toronto-european-gypsy-moth-may-2020

Dunlap, Riley and Curtis Beus. 1992. "Understanding Public Concerns about Pesticide: An Empirical Examination." *Journal of Consumer Affairs* 26(2): 418–438.

Fernandez-Cornejo, Jorge, Craig Osteen, Richard Nehring, and Seth Wechsler. 2014. "Pesticide Use Peaked in 1981, Then Trended Downward, Driven by Technological Innovations and Other Factors." United States Department of Agriculture. Accessed January 29, 2020. https://www.ers.usda.gov/amber-waves/2014/june/pesticide-use-peaked-in-1981-then-trended-downward-driven-by-technological-innovations-and-other-factors/

Fine, Gary Alan. 1988. *Morel Tales: The Culture of Mushrooming.* Cambridge, MA: Cambridge University Press.

Fletcher, Raquel. 2018. "Québec tightens rules on pesticides." *Global News*, February 19.

Foster, John Bellamy. 1999. *The Vulnerable Planet: A Short Economic History of the Environment.* New York: Monthly Review Press.

Frickel, Scott, Sahra Gibbon, Jeff Howard, Joanna Kempner, Gwen Ottinger, and David Hess. 2010. "Undone Science: Charting Social Movement and Civil Society Challenges to Research Agenda Setting." *Science, Technology & Human Values* 35: 444–473.

Galison, Peter. 2008. "Removing Knowledge: The Logic of Modern Censorship." In *Agnotology: The Making and Unmaking of Ignorance*, edited by Robert Proctor and Londa Schiebinger, 37–54. Stanford, CA: Stanford University Press.

Goven, Joanna, Tom Kerns, Romeo Quijano, and Dell Wihongi. 2007. *Report of the March 2006 People's Inquiry Into the Impacts and Effects of Aerial Spraying Pesticide over Urban Areas of Auckland.* Auckland, NZ: Action Plan & Print. https://peoplesinquiry.wordpress.com/

Guivant, Julia. 2003. "Pesticide Use, Risk Perception and Hybrid Knowledge: A Case Study from Southern Brazil." *International Journal of Sociology of Agriculture and Food* 11: 41–51.

Hallberg, George. 1987. "The Impacts of Agricultural Chemicals on Ground Water Quality." *GeoJournal* 15(3): 283–295.

Hannigan, John. 1995. *Environmental Sociology: A Social Constructionist Perspective.* London: Routledge Press.

Hannigan, John. 2014. *Environmental Sociology, 3rd edition*. London: Routledge Press.

Harrison, Jill Lindsey. 2006. "'Accidents' and Invisibilities: Scaled Discourse and the Naturalization of Regulatory Neglect in California's Pesticide Drift Conflict." *Political Geography* 25: 506–529.

Harrison, Jill Lindsey. 2011. *Pesticide Drift and the Pursuit of Environmental Justice*. Cambridge, MA: MIT Press.

Harrison, Jill Lindsey. 2014. "Neoliberal Environmental Justice: Mainstream Ideas of Justice in Political Conflict over Agricultural Pesticides in the United States." *Environmental Politics* 23(4): 650–669.

Healy, David. 2012. *Pharmageddon*. Berkeley: University of California Press.

Henderson, Julie, Annabelle Wilson, Samantha Meyer, John Coveney, Michael Calnan, Dean McCullum, Sue Lloyd, and Paul Ward. 2014. "The Role of the Media in Construction and Presentation of Food Risks." *Health, Risk & Society* 16(7–8): 615–630.

Hess, David. 2007. *Alternative Pathways in Science and Industry: Activism, Innovation, and the Environment in an Era of Globalization*. Cambridge, MA: MIT Press.

Jay, Mairi and Munir Morad. 2006. "The Socioeconomic Dimensions of Biosecurity: The New Zealand Experience." *International Journal of Environmental Studies* 63(3): 293–302.

Johal, Rumneek. 2020. "Here's why you'll see low-flying planes over Surrey next week." *Daily Hive News*, May 1. https://dailyhive.com/vancouver/gypsy-moth-spraying-surrey-2020

Johnson, Stephen, Terre Satterfield, James Flynn, Robin Gregory, C. Mertz, Paul Slovic, and Robert Wagner. 1995. *Vegetation Management in Ontario's Forests: Survey Research of Public and Professional Perspectives*. Toronto, Ontario: Queen's Printer for Ontario.

Jorgenson, Andrew K. and Kennon A. Kuykendall. 2008. "Globalization, Foreign Investment Dependence and Agriculture Production: Pesticide and Fertilizer Use in Less-Developed Countries, 1990–2000." *Social Forces* 87(1): 529–560.

Kasperson, Roger and Jeanne Kasperson. 1996. "The Social Amplification and Attenuation of Risk." *ANNALS of the American Academy of Political and Social Science* 545: 95–105.

Kleinman, Daniel and Sainath Suryanarayanan. 2013. "Dying Bees and the Social Production of Ignorance." *Science, Technology & Human Values* 38(4): 492–517.

Lincer, Jeffrey. 1975. "DDE-induced Eggshell Thinning in the American kestrel: A Comparison of the Field Situation and Laboratory Results." *Journal of Applied Ecology* 12(3): 781.

Lockie, Stewart. 1997. "Chemical Risk and the Self-Calculating Farmer: Diffuse Chemical Use in Australian Broadacre Farming Systems." *Current Sociology* 45(3): 81–97.

Markowitz, Gerald and David Rosner. 2002. *Deceit and Denial: The Deadly Politics of Industrial Pollution*. Berkeley: University of California Press.

Mart, Michelle. 2014. *Pesticides, A Love Story: America's Enduring Embrace of Dangerous Chemicals*. Lawrence: University of Kansas Press.

McCarthy Auriffeille, Deborah and Leslie King. 2014. "Introduction: Environmental Problems Require Social Solutions." In *Environmental Sociology: From Analysis to Action, 3rd edition*, edited by Leslie King and Deborah McCarthy Auriffeille, 1–22. Lanham, MD: Rowman & Littlefield Publishers.

McCulloch, Jock and Geoffrey Tweedale. 2008. *Defending the Indefensible: The Global Asbestos Industry and its Fight for Survival*. Oxford: Oxford University Press.

McGoey, Linsey. 2012. "Strategic Unknowns: Towards a Sociology of Ignorance." *Economy and Society* 41(1): 1–16.

Michaels, David. 2008. *Doubt Is Their Product: How Industry's Assault on Science Threatens Your Health*. New York: Oxford University Press.

Montreal Gazette. 2019. "Glyphosate: Montreal to ban herbicide used in Roundup." *Montreal Gazette*, September 5.

Moyers, Bill, Sherry Jones, Loren Berger, Jackson Frost, and Joseph Camp. 2002. *Trade Secrets: A Moyers Report*. Princeton, NJ: Films for the Humanities and Sciences.

Nader, Laura. 1997. "Controlling Processes: Tracing the Dynamic Components of Power." *Current Anthropology* 38(5): 711–738.

Office of the Ombudsman. 2007. *Report of the Opinion of Ombudsman Mel Smith on Complaints Arising from Aerial Spraying of the Biological Insecticide Foray 48B on the Population of Parts of Auckland and Hamilton to Destroy Incursions of Painted Apple Moths, and Asian Gypsy Moths, Respectively*. Wellington, NZ: Office of the Ombudsman.

Oreskes, Naomi and Erik Conway. 2010. *Merchants of Doubt*. New York: Bloomsbury Press.

Patnaik, Pradyot. 2007. *A Comprehensive Guide to the Hazardous Properties of Chemical Substances*. Hoboken, NJ: John Wiley & Sons.

Pesticide Action Network of North America (PANNA). 2005. "PANNA CDC Body Burden Study Finds Widespread Pesticide Exposure." http://www.panna.org/legacy/panups/panup_20050722.dv.html

Pimental, David and Michael Burgess. 2014. "Environmental and Economic Costs of the Application of Pesticides Primarily in the United States." In *Integrated Pest Management*, Volume 3, edited by David Pimental and Rajinder Peshin, 47–71. New York: Springer Press.

Proctor, Robert. 2008. "Agnotology: A Missing Term to Describe the Cultural Production of Ignorance (and Its Study)." In *Agnotology: The Making and Unmaking of Ignorance*, edited by Robert Proctor and Linda Schiebinger, 1–36. Stanford, CA: Stanford University Press.

Rayner, Steve. 2012. "Uncomfortable Knowledge: The Social Construction of Ignorance in Science and Environmental Policy Discourses." *Economy and Society* 41(1): 107–125.

Reganold, John, Lloyd Elliott, and Yvonne Unger. 1987. "Long-Term Effects of Organic and Conventional Farming on Soil Erosion." *Nature* 330: 370–372.

Robbins, Paul. 2004. "Comparing Invasive Networks: Cultural and Political Biographies of Invasive Species." *Geographical Review* 94(2): 139–156.

Savonen, Carol. 1994. "Btk spraying for forest pest kills many other species." *OSU News*. Corvallis, OR: Oregon State 'University' Agricultural Communications.

Schnaiberg, Alan and Kenneth Gould. 1994. *Environment and Society: The Enduring Conflict*. New York: St. Martin's Press.

Shorette, Kristin. 2012. "Outcomes of Global Environmentalism: Longitudinal and Cross-National Trends in Chemical Pesticide Use." *Social Forces* 91(1): 299–325.

Steingraber, Sandra. 2009. "The Social Construction of Cancer: A Walk Upstream." In *Environmental Sociology: From Analysis to Action*, edited by Leslie King and Deborah McCarthy, 287–299. Lanham, MD: Rowman & Littlefield Publishers.

Sumlin, Todd. 2015. "Cankerworms are back in Charlotte." *The Charlotte Observer*, April 10. https://www.charlotteobserver.com/news/local/article18162221.html

Swadener, Carrie. 1994. "Bacillus Thuringiensis (B.T.)." *Journal of Pesticide Reform* 14(3): 13–20.

Vizcaya, David, Maria Mirabelli, Josep-Maria Antó, Ramon Orriols, Felip Burgos, Lourdes Arjona, and Jan-Paul Zock. 2011. "A Workforce-Based Study of Occupational Exposures and Asthma Symptoms in Cleaning Works." *Occupational and Environmental Medicine* 68(12): 914–919.

Vogel, Sarah. 2012. *Is It Safe? BPA and the Struggle to Define the Safety of Chemicals.* Berkeley: University of California Press.

Walsh, Frances. 2003. "Wipe Out." *Metro* 261: 44–45.

Ward, Neil. 1995. "Technological Change and the Regulation of Pollution from Agricultural Pesticides." *Geoforum* 26(1): 19–33.

Wargo, John. 1996. *Our Children's Toxic Legacy: How Science and Law Fail to Protect Us from Pesticides.* New Haven, CT: Yale University Press.

Washington State Department of Health. 2001. "Report of health surveillance activities: aerial spraying for Asian Gypsy Moth – May 2000 Seattle, WA." https://www.doh.wa.gov/Portals/1/Documents/Pubs/334-292.pdf

West Aucklanders against Aerial Spraying (WASP). 2002. "Helen Wiseman-Dare Speech: Anti-Spray March Nov. 30." December 4. https://www.scoop.co.nz/stories/AK0212/S00023.htm

Wildblood-Crawford, Bruce. 2006. "Grassland Utopia and *Silent Spring*: Rereading the Agrichemical Revolution in New Zealand." *New Zealand Geographer* 62: 65–72.

Woodhouse, Edward and Jeff Howard. 2009. "Stealthy Killers and Governing Mentalities: Chemicals in Consumer Products." In *Killer Commodities: Public Health and the Corporate Production of Harm,* edited by Merrill Singer and Hans Baer, 35–66. Plymouth: AltaMira Press.

Zargar, Darya. 2019. "North Surrey ground zero for aerial spray to rid invasive gypsy moths." *Global News,* May 1. https://globalnews.ca/news/5226512/surrey-gypsy-moth-spray/

1 New Zealand's Painted Apple Moth Eradication Operation

In April 1999 workers at a West Auckland industrial property discovered a population of moths on their site, which prompted the owner to ask Peter Maddison, a local entomologist, to investigate (Panckhurst 2001; Tyson 2009). He identified the moths as painted apple moths (*Teia anartoides*) (PAM), which are a non-indigenous species from Southern Australia, where it is considered a common but minor urban garden pest (Elliott et al. 1998; Suckling et al. 2014; Tyson 2009). After identifying the moths, Maddison contacted the Ministry of Agriculture and Forestry (MAF), which is responsible for dealing with invasive species. They, in turn, carried out an assessment that revealed the moths had spread to 22 properties in West Auckland (Auckland District Health Board 2002). MAF then decided this was a foreign species that needed to be eradicated and began working towards that end.

The first eradication activities included removing host trees and ground spraying infested areas with chlorpyrifos (an organophosphate pesticide) (Auckland District Health Board 2002). However, in October 1999 MAF discovered those activities failed to eliminate the moth as a second population was found in a suburb 15 kilometers away (9.3 miles) (MAF 1999).

MAF's response to the failure was to use a stronger pesticide, which was Decis Forte, a synthetic pyrethroid insecticide containing the active ingredient deltamethrin (Auckland District Health Board 2002; Pesticide Action Network 2010). However, by August 2000, ten months later, the moth still had not been eradicated. This led the Cabinet, in May 2001, to bolster eradication efforts by approving an additional $1.75 million to continue tree host removal and ground spraying with insecticides (Office of the Ombudsman 2007). However, despite the government's commitment to eradication efforts, by October 2001 the moth managed to spread to additional surrounding suburbs, including Avondale, Titirangi, Kelston, Glen Eden and Mt. Wellington (*Ibid.*). The moth's spread was surprising, considering it is less mobile than most moths: while the males can fly, the females cannot, which reduces their mobility to 200–300 meter hops and considerably slows the population's spread (New Zealand Audit Office 2003).

PAM's presence in New Zealand concerned forestry industry officials, as they feared the moth might eventually spread to forestry plantations, which

DOI: 10.4324/9780429426414-2

might lead to tree defoliation that would stunt tree growth and dent profits. Moreover, this concern was exacerbated by MAF's failure to contain, let alone eradicate PAM. Industry officials began communicating their concerns to MAF leadership in November 1999 and continued to do so as the infestation spread (Office of the Controller and Auditor-General 2002). Additionally, by late 2000 industry officials had become so concerned that they began pressing MAF to commission an independent review of the agency's response to the PAM incursion, while also lobbying to escalate the response by adding aerial pesticide spraying of the infested suburbs (*Ibid.*). At the beginning of 2001 MAF acquiesced to industry's demands for an independent review. The reviewers released their report in June 2001, which made 34 recommendations, with one being to start aerial pesticide spraying of the infested areas (*Ibid.*).

Escalating to an Aerial Pesticide Spraying Operation

In October 2001 Helen Clark's Cabinet responded to the moth's spread by giving the Committee for Infrastructure and Environment the power to handle the situation as they saw fit (Office of the Ombudsman 2007). They, in turn, endorsed the proposal for a limited aerial pesticide spraying operation consisting of six to eight sprayings over an area covering 300 ha (741 acres which equates to 420 football fields) in West Auckland (*Ibid.*).

The pesticide chosen for this campaign was Foray 48B, a commercial insecticide containing a live bacterium [*Bacillus thuringiensis subspecies kurstaki* (Btk)] and numerous concealed synthetic chemicals (including, but not limited to, hydrochloric acid, sulfuric acid, propylene glycol, and benzoic acid) that were added to enhance the bacteria's preservation and its ability to bind to plant surfaces.

Bacillus thuringiensis (Bt) is a bacteria that is naturally found in soil and is toxic to certain species. It functions as an insecticide by producing crystal protoxins that impair the digestion of various species, which leads them to starve to death (Rubio-Infante & Moreno-Fierros 2016). It was first commercialized in the late 1930s in France and then in the United States in 1961 (Sanahuja et al. 2011; Siegel 2001). In the subsequent decades its use grew exponentially, and by 1989 it had captured more than 90% of the pesticide market (Feitelson et al. 1992).

Btk is a subspecies of *Bt* that is rarely found in nature, but which is used extensively in agriculture due to its toxicity towards caterpillars (Sanahuja et al. 2011). The bacteria achieves its effects by impairing the digestive system of moth and butterfly larvae. Specifically, when caterpillars eat leaves treated with *Btk*, the bacteria spores in the caterpillar's gut and releases a crystal protoxin (Swadener 1994; Upton & Caspar 2008). As the crystal dissolves in the caterpillar's alkaline gut, it releases delta-endotoxin proteins that bind to the epithelium cells lining the gut and creates pores in the cell membrane, which upsets the gut's ion balance (Rubio-Infante & Moreno-Fierros 2016;

Swadener 1994). The resulting imbalance impairs normal digestion and triggers the insect to stop feeding, which eventually leads it to starve to death (Upton & Caspar 2008). While the bacteria is naturally toxic to caterpillars, pesticide company scientists have engineered strains that are at least six times more potent than is normally found in nature, and these are the strains that are used in aerial spraying operations (Burges & Jones 1998; Sanahuja et al. 2011).

MAF began the aerial spraying operation in January 2002, having doubled the previously proposed spray zone of 300 ha. Also, while they began target spraying infested areas with helicopters, in May 2002, the agency announced it would start using airplanes to blanket spray the pesticide on suburbs where moths had been found (sometimes as few as two moths was all that was required to maintain spraying) (Goven et al. 2007; MAF 2002).

Helen Clark's Labour Government resisted significantly expanding the spraying operation prior to the August 2002 national election. However, after her party secured re-election her cabinet approved $88 million to extend and expand the eradication operation (Office of the Ombudsman 2007). This included adding ten spraying operations over the ensuing eight months and substantially expanding the spray zone, which increased from 722 ha (1,784 acres) in July to 868 ha (2,145 acres) in August, 962 ha (2,377 acres) in September, 7,980 ha (19,719 acres) in October, 8,903 ha (22,000 acres) in December, and 10,632 ha (26,272 acres) in January 2003 (Goven et al. 2007; Office of the Ombudsman 2007).

Although the extended spraying campaign was supposed to end in April 2003, that month government officials announced another spraying would be carried out in May, due to warmer than expected Autumn weather (New Zealand Government 2003a). Then, in May, Jim Sutton (Biosecurity Minister) announced that despite only trapping two moths, they would continue with spraying over the Winter months of June, July, August, and September (New Zealand Government 2003b). Spraying continued during the Spring, Summer, and Autumn months and finally concluded on May 13, 2004 (Goven et al. 2007).

Over the course of the eradication operation MAF carried out aerial spraying operations on over 60 days, which delivered over 703,150 liters (185,753 gallons) of pesticide over West Auckland suburbs (Goven et al. 2007). At its peak the spray area exposed more than 200,000 residents to repeated sprayings, with many experiencing multiple sprayings on the same day (Office of the Ombudsman 2007). Importantly, the 200,000 figure does not account for the transient population, who had to enter the spray zone for work or other reasons (*Ibid.*). Moreover, the 13,500 people living near the infestation "hot" zone were subjected to nearly 40 sprayings, in addition to the ground spraying operations with deltamethrin (Blackmore 2020).

The operation's final overall cost was estimated to have reached $65 million NZD (Office of the Ombudsman 2007). However, that sum was only the cost of applying the pesticides and providing basic healthcare services related to the spraying (which only consisted of a telephone healthline and diagnostic services). The sum failed to consider the uncalculated damage to local ecosystems, which, as I detail below, was likely to have been considerable. Nor did it reflect the costs residents had to bear for treating the medical problems (both physical and psychological) resulting from exposure to the pesticide spraying, nor the costs of reduced productivity or enjoyment of life resulting from the medical problems. Nor did the sum reflect the other costs that residents had to assume, which included evacuation costs, or the impact of spraying on gardens, clothing, homes, and cars.

Ecological Concerns about the Pesticide Spraying Operation

Although the spraying campaign eradicated PAM, there are many concerns about the impact it likely had on local ecosystems. First, it is likely to have significantly harmed populations of indigenous moths and butterflies due to *Btk*'s indiscriminate effects on *all* caterpillars (Upton & Caspar 2008). The U.S. Environmental Protection Agency (1998) reported that *Bt* is harmful to Lepidopteran species (i.e. the class of insects including moths and butterflies), which corresponds with several field reports. For instance, one Oregon *Btk* spraying operation reduced the total weight of local caterpillars by 90%–95%, the number of caterpillars by 80% and the number of caterpillar species by 60% (Savonen 1994). Similarly, a spraying to control spruce budworm in Oregon led to lower numbers of caterpillars (Miller 1990b). In a third case researchers found that a single treatment of *Btk* led to an 80% decrease in Lepidoptera larvae, compared to the pre-spraying sample (Miller 1990b).

It is important to note these effects often persisted well beyond the spraying event. For example, researchers have found that *Btk* droplets remain lethal to the swallowtail butterflies 30 days after its application (Johnson et al. 1995; Miller & West 1987). In another case, while Carol Savonen (1994) found that a *Btk* spraying led to a 80% decrease in indigenous caterpillars, she also found that the following year there were 71% fewer larvae than prior to the spraying. In contrast, she found untreated areas had 30% more species and five times the numbers of caterpillars (*Ibid.*). As well, applications of Foray 48B (which has *Btk* as its active ingredient) in West Virginia reduced the number of caterpillar species and total caterpillar numbers, with the numbers remaining reduced a year later (Swadener 1994). In still another case, Miller (1990a) found that after administering a *Bt* spray in Oregon, to kill gypsy moth larvae, the number of indigenous caterpillars was reduced for two years. Moreover, the number of oak-feeding caterpillars, in particular, was reduced for three years (*Ibid.*).

Importantly, the New Zealand government was well aware of the potential for such effects as their Environmental Impact Report warned that:

> Any business involved in rearing moths and butterflies commercially, for example butterfly farms, would carry a heavy risk of damage if located within or near a spray area. Scientific institutions with insect rearing facilities for biological control work or study of Lepidoptera species would

similarly be at risk. School biology classes or private individuals rearing caterpillars, such as monarch butterflies on swan plants, could expect caterpillar mortality if they were exposed to the spray.

(MAF 2003)

A second ecological concern is the spray's likely impact on other non-caterpillar insects as post-spraying studies have confirmed that using *Bt*-based sprays decreases numerous beneficial insects. For example, Salama et al. (1991) reported that the use of *Bt Berliner* impacted parasitic wasps, which included reducing the number of eggs they deposited and the percentage of eggs hatched. Additionally, there are reports that using Dipel (a commercial formulation that has *Btk* as its active ingredient) kills predatory mites, cinnabar moths, and aphid-eating flies, each of which contributes natural biological control services (Chapman & Hoy 1991; Horn 1983; James et al. 1993). As well, British Columbia's 1993–1994 spraying of Foray 48B was followed by a reduction in honeybee, bumblebee, ladybug, and skipper populations (Young 1994).

Such insect losses are concerning as they diminish the ecosystem services these insects provide, such as the pollination services provided by bees, butterflies, and other pollinating species. Another ecosystem service they provide is biological control of unwanted species. For example, parasitic wasps are important for providing natural population control of unwanted insects, such as aphids, corn borers, planthoppers, whiteflies, and wood borers (Wang et al. 2019). Interestingly, there were reports that parasitic wasps had infected many of the female painted apple moths that were part of the first detected population in Glendene, which suggests indigenous species had already started working to control the PAM population (WASP 2002b). Beyond pollination and biological control, insects also provide weed control, decomposition, as well as serving as a food source for birds, bats, and other beneficial species (Jankielsohn 2018; Swadener 1994).

Although New Zealand researchers have done a poor job of tracking the economic benefits provided by insects, researchers in other countries have identified their considerable contributions. For example, Losey and Vaughan (2006) estimated that in the United States the economic value of insect ecosystem services surpasses $57 billion per year. As well, Carreck and Williams (1998) estimated that, in the United Kingdom, bees produce over £200 million (~$322 million USD in December 1999 (Pound Sterling Live n.d.)) of value each year, with the bulk (i.e. £137.8 million) coming from pollination services.

Beyond harming beneficial insects, *Bt* sprays can also harm higher-order species, such as fish and birds. Regarding the former, there are reports that high concentrations of Foray 48B are acutely toxic to rainbow trout, with other researchers finding similar results for juvenile coho salmon (Surgeoner & Farkas 1990, as cited by SES 2003). As for birds, a study on the effects of Dipel (which also contains *Btk*) found that egg hatching of pheasant eggs were only half as successful as untreated populations (Jones 1986, as cited by Swadener 1994). In another case, following a spraying of Foray 48B in urban areas of British Columbia, there were reports of dead fledgling birds, chickens unexpectedly dying, birds producing black feces, and whole chicken flocks coming down with diarrhea (Young 1994). As well, there were many reports of wild and pet bird deaths following 2002–2004 spraying events against the painted apple moth (Blackmore 2003).

Aside from appearing to harm some bird species, *Bt* applications also impact birds by reducing their food source. For example, following a *Bt* treatment in New Hampshire, researchers noted blue-throated warblers made fewer nesting attempts and brought fewer caterpillars to their nests (Rodenhouse & Holmes 1992). Other research found that *Btk*-treated areas had fewer spruce grouse chicks and that those chicks developed more slowly than in untreated areas (Norton et al. 2001). In another case the caterpillar reductions brought on by *Btk* spraying reduced black-throated blue warbler breeding below the number needed to balance out annual mortality levels (Rodenhouse & Holmes 1992).

There are also concerns for mammals as there is evidence *Bt* bacteria can persist in the bodies of mammals. First, although it has long been believed *Bt* bacteria cannot survive a mammal's acidic gut environment, there is scientific evidence suggesting otherwise. For instance, Jensen et al. (2002) found that *Btk* spores can germinate in the intestines of humans (Jensen et al. 2002). Besides the gut, there is also evidence the bacteria can persist in other organs. For example, Tsai et al. (1997) found *Bt* bacteria persisted 21 days in the lungs of rats, while Siegel et al. (1987) found it persisted in the lungs of mice for 2.5 days after respiratory exposure. Additionally, Siegel (2001) found that *Btk* persisted in the spleen of mice 37 days after intraperitoneal injection.

In addition to persisting in mammals, there are reports the bacteria can cause harm. For instance, in one study rabbits exposed to *Bt* experienced irritation to the skin and eyes (Novo Nordisk 1991, as cited in Swadener 1994; Seigel et al. 1987). As well, sheep exposed to *Btk* through diet had loose stools, with some displaying microscopic damage to their colons (Hadley et al. 1987). Additionally, Hernandez et al. (1998) reported a 80% mortality rate for mice experiencing intranasal exposure to high concentrations of *Btk* spores. What's more, during the PAM spraying campaign there were over 20 reports of cats being unwell following sprayings, with typical symptoms including vomiting, lack of hunger, infected eyes, and skin allergies (Blackmore 2003). There were also reports of dogs being affected, with the most common symptom being diarrhea, followed by eye and skin problems (*Ibid.*). A reason animals are being so impacted is that *Btk* is not something they would be typically exposed to in the environment, as *Btk* is a relatively rare strain of Bt. Moreover, as previously stated, the *Btk* strains found in commercial pesticides have been engineered to be up to six times more potent than those typically found in nature (Sanahuja et al. 2011; Swadener 1994).

Another ecological concern with *Btk* is its persistence in the environment. While MAF originally assured residents the spray's active ingredient would break down after two hours, MAF scientists subsequently found it can persist up to 11 days (WASP 2002a). The latter is consistent with previous research that found the insecticide's half-life can exceed nine days (West & Burges 1985). Moreover, some studies suggest Bt spores can persist in the environment even longer. In one case viable *Btk* spores were recovered from tree foliage a full year after application (Feitelson et al. 1992). Moreover, in Denmark *Btk* was found in a cabbage field seven years after application (Van Cuyk et al. 2011). Similarly, Vettori and colleagues (2003) found that *Btk* bacteria survived 88 months after spraying and that the bacteria's toxin persisted for 28 months after spraying.

A related problem is persistence in water. In some cases *Btk* has been detected in the rivers and public water systems after aerial spraying, and it has been found that *Btk* spores are not adequately destroyed by standard water treatment processes (Menon & Mestral 1985). In another experiment, researchers detected viable *Bt* cells in the water for up to 200 days after application and in sediment for up to 270 days (Hoti & Balaraman 1991).

Still another ecological concern is potential contamination with genetically modified organisms (GMO) as the spray's bacterial ingredient was grown on a nutrient broth consisting largely of soy and corn from the United States, where virtually all supplies are GMO-contaminated (Pesticide Action Network Aotearoa 2002). While MAF confirmed the spray contained soy and corn, they did not test it for GMO contamination, which strongly suggests Western Auckland was thoroughly doused with GMO contaminants, thereby undermining attempts at organic agriculture in the area. This should have been particularly concerning in light of the country's previous strong opposition to genetically engineered organisms and the government's efforts to portray the country as "100% Pure and Natural" to the rest of the world (Rudzitis & Bird 2011)

Human Health Concerns with the Pesticide Spraying

The PAM eradication campaign introduced numerous concerns for human health. First, the pesticides used for the initial ground spraying contained chlorpyrifos, an organophosphate pesticide that can lead to developmental problems, particularly in children whose mothers were exposed during pregnancy (Rauh et al. 2011). As well, the chemical is associated with (1) severe neurological effects; (2) reproductive harms; (3) metabolic effects (such as obesity and diabetes); (4) respiratory problems, including asthma; and (5) childhood cancers (Californians for Pesticide Reform 2007; CHE 2020; Rauh et al. 2011).

The second pesticide that was deployed was deltamethrin, which has been found to impact the neurology, liver, heart, kidneys, immune systems, and sperm development of lab animals (Haverinen & Vornanen 2016; Khalatbary et al. 2016; Kumar et al. 2015; Kumar & Sharma 2015; Sharma et al. 2014; Tos-Luty et al. 2001; Tuzmen et al. 2008). Moreover, in humans it has been known to cause dermal issues (such as burning, numbness, and tingling), headaches, dizziness, fatigue, nausea, and anorexia (Barlow et al. 2001).

Concerns about the Aerial Pesticide (i.e. Foray 48B)

Regarding the aerial pesticide spraying, while the government repeatedly reassured the public that Foray 48B was completely safe, there is much data suggesting otherwise (Office of the Ombudsman 2007).

First, there was documented evidence that previous uses of Foray 48B were associated with a range of health problems, including respiratory, digestive, and neurological ailments. For instance, ground spray applicators in Vancouver (Canada) reported eye, nose, throat, and respiratory irritation (Noble et al. 1992). Health complaints also emerged in urban areas where Foray 48B was sprayed. According to the Washington State Health Department 250 people reported health problems following the 1993 spraying over Spokane, with another 59 reporting health problems following the 2000 spraying operation over Seattle (Washington State Department of Health 1993, 2001).

Health complaints were also reported in New Zealand, starting with the 1996–97 spraying operation against the white-spotted tussock moth (which also used Foray 48B) in East Auckland, where a community health monitoring program identified 375 residents who suffered health problems attributed to the spraying (Office of the Ombudsman 2007). The reported symptoms included respiratory symptoms (including asthma, chest tightness, and coughs), headaches, skin irritation, skin rashes, sore throats, blocked noses, eye irritation, diarrhea, vomiting, stomach cramps, flu-like symptoms (such as fever, malaise, and swollen glands), and lethargy (Hales et al. 2004; Office of the Ombudsman 2007; Watts 2003). Additionally, a Ministry of Forestry study revealed that 8% of East Auckland residents reported being affected by that spraying program, that the figure went up to 9.9% in the more frequently sprayed areas, and 16.7% for those residing in the highest sprayed "hot zones" (Auckland Healthcare 1997; Allpro consulting 1997, as cited by Blackmore 2020).

Substantial evidence also emerged over the course of the PAM eradication operation's first year. First, prior to the expansion of the spraying operation, researchers from the University of Auckland's Faculty of Medical and Health Sciences carried out a door-to-door survey in the 500 hectare core of the spraying zone, which assessed symptom complaints ten weeks before and ten weeks after spraying (Petrie et al. 2003). They found a significant increase in symptom complaints, and the reported symptomology included respiratory problems, gastrointestinal issues, and neuropsychiatric symptoms. Second, a large pharmaceutical company (i.e. Douglas Manufacturing) located in the spray zone conducted a survey of its employees and found that 15% of them experienced health effects from the spraying or had family members who had (Blackmore 2003). Reported symptoms included, in order of frequency, eye problems (itchy, watery, or sore eyes), lung and/or respiratory problems (breathing difficulty, asthma attack, respiratory irritation, nosebleeds, sinus pain, and sneezing), skin burning, nausea or upset stomach, and headaches (*Ibid.*). Further evidence emerged shortly after the spraying operation's first year, as a community health monitoring program recorded 315 residents who had experienced respiratory, digestive, and/or neurological problems during spraying, all consistent with health issues that emerged during the 1996–1997 spraying operation in East Auckland and from other reports (Blackmore 2003).

Later in 2003 the New Zealand Education Institute (NZEI 2003) conducted a survey of West Auckland primary schools and found that 56% of the 320 responding staff reported adverse health effects from the spraying. The symptoms they reported included "rashes, nausea, persistent coughing, breathlessness, asthma attacks, mucous in nose and throat, tightness in chest, difficulty breathing, allergic (histamine) reactions, swollen and sore throats, sores around moth, watery eyes and bronchial problems" (NZEI 2003: 1). As well, many reported seeing similar symptoms among the children under their care (*Ibid.*).

Evidence of negative impacts also emerged from government sources as MAF's Health Service fielded 22,643 calls about health problems related to the spraying and provided clinical assessments to 840 people, with 100 of those being referred on for specialist consultation (Office of the Ombudsman 2007). Moreover, the Ombudsman (2007) reports approximately 3,500 people reported illness from the spraying. Relatedly, there were still more than 3,600 people registered with the PAM Health Service in March 2004, which was 27 months after spraying had begun (Blackmore 2020). The list of symptoms reported to MAF included fever, swollen glands, skin rashes, diarrhea, vomiting, stomach cramps, coughing, asthma, congested nose, headaches, as well as sore eyes or throat (Blackmore 2003; Hales et al. 2004).

It bears noting that the numbers cited above are very likely to be an under-reporting, as health workers typically overlook the chronic effects of pesticide exposure (Solomon 2000). Moreover, this is more likely to have been true in this case, given that the medical service was being provided by a private firm hired by MAF, who was thus inclined to do MAF's bidding. As Barbara Ellen Smith (1981) has shown, medical practice is strongly influenced by funding arrangements, with those who pay the piper getting to call the tune. This is a point I revisit in Chapter 8. Another point worth mentioning is that the aforementioned numbers would have been higher but for the fact that prior to each spraying the government evacuated 694 people with pre-existing health issues (Blackmore 2020).

Further evidence of harm emerged during the eight-week aerial spraying operation MAF conducted in late 2003, over the city of Hamilton, in order to eradicate the Asian Gypsy Moth. Out of the 24,000 people who were in the spray zone, 855 people reported illness due to the spraying, which represents 3.6% of the population (Office of the Ombudsman 2007). Additionally, a community group collected and collated self-reported health effects from 202 people, who reported the following health problems: respiratory problems (233), neurological complaints (139), skin irritations (79), eye infections (67), digestive problems (61), as well as fatigue and other symptoms (46).

Another source of evidence comes from Occupational Safety and Health (OSH), which investigated complaints from staff at Fraser High School (in Hamilton) who experienced severe health reactions following the first two sprayings and were being dismissed by Aeraqua, the medical professionals the government hired to provide residents with health services during the Auckland and Hamilton spraying operations (OSH 2003, as cited by Blackmore 2020). OSH reported that over 30 staff were negatively affected by the spraying and that many suffered serious reactions (*Ibid.*). As well, the agency concluded that "A causal link between adverse health effects and occupational exposure to Foray 48B has been established in a number of staff members, a number of whom had rare food allergies" (OSH 2003, as cited in Office of the Ombudsman 2007: 64).

There is also evidence linking the spraying to asthma. For instance, Gallagher and colleagues (2005) found that during the 2002–2004 period there was a doubling in the asthma discharge rate for boys aged 0–4 who were in the spray zone and a 50% increase for girls of the same age. Similarly, Hales and colleagues (2005) found a significant space-time clustering of West Auckland children who were admitted for acute asthma during the spraying operations. Beyond identifying an increase in cases, they found that the cases clustered within the boundaries of the aerial spray zone.

Furthermore, while government officials assured the pesticide was safe during the spraying operations in Auckland and Hamilton, in 2006, two years after completing the PAM operation, the government finally acknowledged that Foray 48B was associated with health problems, including physical symptoms (i.e., flu-like symptoms, stomach discomfort, diarrhea, chest tightness, and irritation of the throat, nose, eyes, and skin) and neuropsychiatric problems (i.e., anxiety, dizziness, sleep problems, and concentration difficulties) (Frampton et al. 2006, as cited by Office of the Ombudsman 2007).

Concerns about the Main Ingredient (Btk)

The pesticide's main ingredient is the *Btk* bacteria, and while government officials consistently reassured citizens it was safe to humans, there is evidence suggesting this claim was also suspect.

One problem is the bacteria's capacity to persist in humans, as underscored in monitoring studies. For instance, a monitoring program undertaken after a Vancouver (Canada) spraying found that 11% nasal swabs taken from patients contained *Btk* (Noble et al. 1992). Additionally, nearly all workers tested positive for *Btk* if they had been repeatedly exposed (5–20 times) to high concentrations, and most remained culture positive for 14 to 30 days (*Ibid.*).

There is also evidence *Btk* spores can germinate in human intestines, which contradicts the commonly held belief that *Bt* bacteria cannot survive in mammals' acidic digestive tracts (Jensen et al. 2002). Relatedly, some researchers have found that the germination of *Btk* strains produces a toxin that causes food poisoning symptoms, including nausea and vomiting (Damgaard, 1995;

Tayabali & Seligy, 2000). Although the issues are particularly concerning for those with impaired immune systems, some studies have shown that *Btk* spores can also persist in humans with healthy immune systems (Valadares et al. 2001).

A more subtle indicator of harm are immune responses that have been detected among those exposed to *Btk*. Bernstein et al. (1999) found that farm workers who had been exposed to commercial *Btk* (as opposed to the form found in nature) had developed immune responses to the bacteria (i.e. the formation of IgG and IgE antibodies to vegetative *Bt* extracts) and that this was particularly true for workers with high exposures. In a similar study, Doekes and colleagues (2004) found that 25% of greenhouse workers exposed to *Btk*-manifested immune responses, with some responses lasting up to three years, which suggests the bodies are fighting off a reproducing population.

Third, there is much to suggest inhaling *Btk* can cause health problems, starting with the manufacturer's 1991 Material Safety Data Sheet for Foray 48B, which states "Repeated exposure via inhalation can result in sensitization and allergic response in hypersensitive individuals" (Novo Nordisk 1991, as cited in Swadener 1994). There are also several studies that have found high Btk concentrations can lead to human health problems (Bernstein et al. 1999; Doekes et al. 2004; Noble et al. 1992). For instance, Noble et al. (1992) found that two-thirds of occupationally exposed spray workers developed symptoms, which included dry skin, headache, chapped lips, as well as transient irritation of eyes, nose, and throat. By contrast, they only found such symptoms in one-third of the workers who were not exposed to the spray. Beyond the research on humans, the aforementioned effects on mice, rats, rabbits, sheep, dogs, and cats suggest systemic effects across mammals, which should raise further concerns for human safety.

Although there is some research suggesting *Bt* sprays are safe to humans (de Amorim et al. 2001; Pearce et al. 2002), such studies have significant methodological issues that limit their generalizability to the PAM eradication operation. In particular, Simon Hales (2004) argues that such epidemiological studies are limited by "subjective or potentially biased assessment of health effects, potential or actual exposure of control groups, and limited duration of follow-up" (p. 401). Furthermore, he argues they are limited in their ability to "detect effects that occur in a small proportion of exposed people" and, therefore, "do not provide strong evidence in support of the long-term safety of *Bt* products in a community setting" (*Ibid.*).

The salience of those studies is further weakened by the fact the populations in those studies experienced far less exposure to the pesticide than was the case during the Auckland spraying campaign. For example, where 13,500 Aucklanders living near the "hot" infestation zone were initially exposed to six to eight sprayings and eventually 40+ sprayings over the course of the spraying operation, the participants in the Pearce et al. (2002) and de Amorim et al. (2001) studies only experienced three sprayings each.

Concerns about the Synthetic Ingredients

Concerns also existed about the synthetic chemicals in Foray 48B. While the New Zealand government concealed the identity of those ingredients from the public, it is known that previous Foray 48B formulations have contained sodium hydroxide, phosphoric acid, sulfuric acid, potassium phosphate, phosphine, and methyl paraben (Swadener 1994). Moreover, when Auckland activists made Freedom of Information act requests they were able to identify three more synthetic chemicals in the Foray 48B formulation: hydrochloric acid, propylene glycol, and benzoic acid (Kedgley 2003).

Lab studies have shown that exposure to each of those three chemicals can, by themselves, impact mammals. For example, an inhalation exposure study found that rats exposed to benzoates displayed lung and trachea irritation, decreased kidney weight, decreased liver weight, and significant death rates (1 out of 6) at the highest exposure rates (Velsicol Chemical Corp. 1981, as cited by Wibbertmann et al. 2000). These results should be concerning as rats are used to model the impact of chemicals on other mammals, including humans. Regarding humans, while benzoic acid is considered safe as a food preservative, being exposed to an aerosol form causes asthma, and irritates the skin and eyes (Kedgley 2003).

Additionally, the inhalation of hydrochloric acid fumes is known to cause choking and inflammation of the respiratory tract (*Ibid.*), with medical research also linking it to asthma (Vizcaya et al. 2011), lung cancer (CHE 2016a), and bronchitis (Brunekeef & Holgate 2002; CHE 2016a). Moreover, propylene glycol has been linked to hearing loss, skin irritation, intestinal damage, and depression and has been found to affect children's central nervous systems (CHE 2016b; Kedgley 2003).

Beyond those chemicals, public-interest scientists uncovered four other chemicals from previous 48B formulations that can cause health effects, three of which (sulfuric acid, phosphine, and sodium hydroxide) cause serious problems. Phosphoric acid has the mildest effects, which include irritating skin and mucous membranes, and having its vapors cause coughing and throat irritation (Patnaik 2007). Sodium hydroxide, also known as lye, is severely corrosive to eyes, skin, mucous membranes, and digestive systems (Harte et al. 1991). Moreover, breathing sodium hydroxide dust or mist irritates the nose's mucous membranes, which can damage the upper respiratory tract in more severe cases (*Ibid.*). Sulfuric acid is a very corrosive liquid that can cause severe skin burns and permanent vision loss (Patnaik 2007). As well, inhaling it as a mist or vapor can produce coughing, significant bronchial constriction, and bronchitis (Patnaik 2007; Swadener 1994). Moreover, chronic exposure can produce bronchitis, conjunctivitis, skin lesions, and erosion of teeth (*Ibid.*). As for phosphine, it is related to hepatitis, pulmonary edema, and seizures (CHE 2016c). As well, there is preliminary evidence linking phosphine to cataracts, chronic renal disease, heart attacks, and peripheral neuropathy (*Ibid.*).

Furthermore, while much has been written about *Btk*, little has been written about the short- and long-term effects of exposing humans and ecosystems to each of the inert chemicals in Foray 48B (Upton & Caspar 2008). Nor has anyone examined the short- and long-term synergistic effects resulting from being exposed to that combination of chemicals (*Ibid.*).

Summary

All pesticide use is associated with environmental and human health concerns. As this chapter has demonstrated, the same was true with the pesticides used during the PAM eradication campaign, which were associated with numerous ecological and human health concerns. Beyond the 2002–2004 Auckland operation, this chapter highlights significant potential problems associated with urban pesticide spraying operations that are still regularly occurring in Canada and the United States.

Despite the cited problems, the New Zealand government proceeded with its spraying operation and even expanded it over time. Subsequent chapters will elucidate the social processes that led to this outcome, with the next one illuminating the social processes that contributed to PAM's arrival in New Zealand and its eventual spread beyond its original point of arrival. Thereafter, the focus will pivot to the processes that contributed to the government's decision to eradicate the species and to do so with pesticides.

References

Auckland District Health Board, Public Health Service. 2002. *Health Risk Assessment of the 2002 Aerial Spray Eradication Programme for the Painted Apple Moth in Some Western Suburbs of Auckland.* Auckland, New Zealand: Auckland District Health Board.

Barlow, Susan M., Frank M. Sullivan, and Jo Lines. 2001. "Risk Assessment of the Use of Deltamethrin on Bednets for the Prevention of Malaria." *Food and Chemical Toxicology* 39: 407–422.

Bernstein, Leonard, Jonathan Bernstein, Mauren Miller, Sylvia Tierzieva, David Bernstein, Zana Lummus, MaryJane Selgrade, Donald Doerfler, and Verner Seligy. 1999. "Immune Responses in Farm Workers after Exposure to *Bacillus thuringiensis* Pesticides." *Environmental Health Perspectives* 107(7): 575–582.

Blackmore, Hana. 2003. *Interim Report of the Community-based Health and Incident Monitoring of the Aerial Spray Programme: January-December 2002.* Auckland February 2003.

Blackmore, Hana. 2020. *Btk Pesticide Sprays: Adverse Health Impacts and Effects: Cover up or Abdication of Responsibility?* Unpublished report. Auckland, New Zealand.

Brunekeef, Bert and Stephen T. Holgate. 2002. "Air Pollution and Health." *Lancet* 360(9341): 1233–1242.

Burges, H. Denis and Keith A. Jones. 1998. "Trends in Formulations of Microorganisms and Future Research Requirements." In *Formulation of Microbial Biopesticides: Beneficial Microorganisms, Nematodes and Seed Treatments*, edited by H. Denis Burges, 311–332. Dordrecht, Netherlands: Springer Netherlands.

Californians for Pesticide Reform. 2007. "Airborne Poisons: Pesticides in Our Air and in Our Bodies." Accessed March 4, 2016. http://pesticidereform.org/downloads/Biodrift-Summary-Eng.pdf

Carreck, Norman and Ingrid Williams. 1998. "The Economic Value of Bees in the UK." *Bee World* 79(3): 115–123.

Chapman, Mary and Marjorie Hoy. 1991. "Relative Toxicity of *Bacillus thuringiensis* var. tenebrionis to the Two-Spotted Spider Mite (*Tetranychus urticae* Koch) and Its Predator *Measeiulus occidentalis* (Nesbitt) (Acari, Tetranychidae and Phytoseiidae)." *Journal of Applied Entomology* 111: 147–154.

Collaborative on Health and the Environment (CHE). 2016a. "CHE Toxicant and Disease Database Entry for 'hydrochloric acid'." Accessed April 3, 2016. http://www.healthandenvironment.org/tddb/contam/?itemid=2651

Collaborative on Health and the Environment (CHE). 2016b. "CHE Toxicant and Disease Database Entry for 'propylene glycol'." Accessed April 3, 2016. http://www.healthandenvironment.org/tddb/contam/?itemid=2995

Collaborative on Health and the Environment (CHE). 2016c. "CHE Toxicant and Disease Database Entry for 'phosphine'. Accessed April 3, 2016. http://www.healthandenvironment.org/tddb/contam/?itemid=2364

Collaborative on Health and the Environment (CHE). 2020. "CHE Toxicant and Disease Database Entry for 'chlorpyrifos'. Accessed April 3, 2016. http://www.healthandenvironment.org/tddb/contam/?itemid=2364

Damgaard, Per Hyldebrink. 1995. "Diarrhoeal Enterotoxin Production by Strains of *Bacillus thuringiensis* Isolated from Commercial *Bacillus thuringiensis*-based Pesticides." *FEMS Immunology and Medical Microbiology* 12: 245–250.

de Amorim, Giovana, Beatrixe Whittome, Benjamin Shore, and David B. Levin. 2001. "Identification of *Bacillus thuringiensis* Subspecies *Kurstaki* Strain HD1-like Bacteria from Environmental and Human Samples after Aerial Spraying of Victoria, British Columbia, Canada with Foray 48B." *Applied Environmental Microbiology* 67: 1035–1043.

Doekes, Gert, Preben Larsen, Torben Sigsgaard, and Jesper Baelum. 2004. "IgE Sensitization to Bacterial and Fungal Biopesticides in a Cohort of Danish Greenhouse Workers: The BIOGART Study." *American Journal of Industrial Medicine* 46(4): 404–407.

Elliott, Humphrey, Cliff Ohmart, and Ross Wylie. 1998. *Insect Pests of Australian Forests: Ecology and Management.* Chatswood, NSW: Reed International Books.

Feitelson, Jerald S., Jewel Payne, and Leo Kim. 1992. "*Bacillus thuringiensis*: Insects and Beyond." *Bio/Technology* 10: 271–275.

Frampton, Ruth, J. D. Stark, C. M. Frampton, T. R. Glare, and L. Beckert. 2006. "Environmental and Health Impacts of Aerially-Applied Btk-based Insecticides." Ministry of Agriculture and Forestry document. Wellington, New Zealand.

Goven, Joanna, Tom Kerns, Romeo Quijano, and Dell Wihongi. 2007. *Report of the March 2006 People's Inquiry Into the Impacts and Effects of Aerial Spraying Pesticide over Urban Areas of Auckland.* Auckland, NZ: Action Plan & Print. https://peoplesinquiry.wordpress.com/

Hadley William M., Scott W. Burchiel, Thomas D. McDowell, John P. Thilsted, Clair M. Hibbs, Jerry A. Whorton, Phillip W. Day, Mitchell B. Friedman, and Raymond E. Stoll. 1987. "Five-Month Oral (Diet) Toxicity/Infectivity Study of *Bacillus thuringiensis* Insecticides in Sheep." *Fundamental and Applied Toxicology* 8(2): 236–242.

Hales, Simon, Virginia Baker, Kevin Dew, Losa Moata'ane, Jennifer Martin, Tim Rochford, David Slaney, and Alistair Woodward. 2004. *Assessment of the Potential Health Impacts of the 'Painted Apple Moth' Aerial Spraying Programme, Auckland.* For the New Zealand Ministry of Health.

Hales, Simon, Clive E. Sabel, Daniel J. Exeter, Julian Crane, and Alistair Woodward. 2005. *Clustering of Childhood Asthma Hospital Admissions in New Zealand, 1999–2004.* Conference paper delivered at SIRC 2005 the 17th Annual Colloquium of the Spatial Information Research Centre University of Otago, Dunedin, New Zealand. https://ourarchive.otago.ac.nz/handle/10523/743

Harte, John, Cheryl Holden, Richard Schneider, and Christine Shirley. 1991. *Toxics A to Z: A Guide to Everyday Pollution Hazards.* Berkeley: University of California Press.

Haverinen, Jaakko and Matti Vornanen. 2016. "Deltamethrin Is Toxic to the Fish (Crucian Carp, Carassius carassius) Heart." *Pesticide Biochemistry and Physiology* 129: 36–42.

Hernandez, Eric, Francoise Ramisse, Jean-Pierre Ducoureau, Thierry Cruel, and Jean-Didier Cavallo. 1998. "*Bacillus thuringiensis subsp. konkukian* (serotype H34) Superinfection: Case Report and Experimental Evidence of Pathogenicity in Immunosuppressed Mice." *Journal of Clinical Microbiology* 36(7): 2138–2139.

Hoti, S. and K. Balaraman. 1991. "Changes in the Populations of *Bacillus thuriengiensis* H-14 and *Bacillus spasacus* Applied to Vector Breeding Sites." *The Environnmentalist* 11(1): 39–44.

Horn, Dave. 1983. "Selective Mortality of Parasitoids and Predators of *Myzus persicae* on Collards Treated with Malathion, Carbaryl, or *Bacillus thuringiensis*." *Entomologia Experimentalis et Applicata* 34: 208–211.

James, Rosalind, Jeffrey Miller, and Bruce Lighthart. 1993. "*Bacillus thuringiensis* var. kurstaki Affects a Beneficial Insect, the Cinnabar Moth (Lepidoptera: Arctiidae)." *Journal of Economic Entomology* 86(2): 334–339.

Jankielsohn, Astrid. 2018. "The Importance of Insects in Agricultural Ecosystems." *Advances in Entomology* 6: 62–73.

Jensen, Gert, Preben Larsen, Bodil Jacobsen, Bodil Madsen, Lasse Smidt, and Lars Andrup. 2002. "*Bacillus thuringiensis* in Fecal Samples from Greenhouse Workers after Exposure to *B. thuringiensis*-Based Pesticides." *Applied and Environmental Microbiology* 68(10): 4900–4905.

Johnson, Kelly S., J. Mark Scriber, James K. Nitao, and David R. Smitley. 1995. "Toxicity of *Bacillus thurigiensis* var. *kurstaki* to Three Non-Target Lepidoptera in Field Studies." *Environmental Entomology* 24(2): 288–297.

Jones, I. W. 1986. *Summary Report: Effect of Dipel and Plyac. on Hatchability of Ringneck Pheasant Eggs.* Oregon Department of Fish and Wildlife.

Khalatbary, Ali Rezakhalatbary, Hassan Ahmadvand, Davood N. Z. Ghabaee, Abbasali K. Malekshah, and Azam Navazesh. 2016. "Virgin Olive Oil Ameliorates Deltamethrin-Induced Nephrotoxicity in Mice: A Biochemical and Immunohistochemical Assessment." *Toxicology Reports* 3: 584–590.

Kedgley, Sue. 2003. "Toxic Ingredients of Painted Apple Moth Spray Tabled." Accessed November 15, 2015. http://www.greens.org.nz/speeches/toxic-ingredients-painted-apple-moth-spray-tabled

Kumar, Anoop, D. Sasmal, and Neelima Sharma. 2015. "Immunomodulatory Role of Piperine in Deltamethrin Induced Thymic Apoptosis and Altered Immune Functions." *Environmental Toxicology and Pharmacology* 39(2): 504–514.

Kumar, Anoop and Neelima Sharma. 2015. "Comparative Efficacy of Piperine and Curcumin in Deltamethrin Induced Splenic Apoptosis and Altered Immune Functions." *Pesticide Biochemistry and Physiology* 119(March): 16–27.

Losey, John and Macey Vaughan. 2006. "The Economic Value of Ecological Services Provided by Insects." *BioScience* 56(4): 311–323.

Menon, A. S. and Jacqueline de Mestral. 1985. "Survival of *Bacillus thuringiensis* var. *kurstaki* in Waters." *Water, Air, and Soil Pollution* 25(3): 265–274.

Miller, Jeffrey. 1990a. "Field Assessment of the Effects of a Microbial Pest Control Agent on Nontarget Lepidoptera." *American Entomology* (Summer): 135–139.

Miller, Jeffrey. 1990b. "Effects of Microbial Insecticide, *Bacillus thuringiensis kurstaki*, on Nontarget Lepidoptera in a Spruce Budworm-Infested Forest." *Journal of Research on the Lepidoptera* 29(4): 267–276.

Miller, Jeffrey and Kennth J. West. 1987. "Efficacy of *Bacillus thuringiensis* and Diflubenzeron on Douglas-fir and Oak for Gypsy Moth Control in Oregon." *Journal of Aboriculture* 13(10): 240–242.

Ministry of Agriculture and Forestry (MAF). 1999. "MAF Seeks Public Assistance in Moth Search." *Scoop Independent News*, October 5.

Ministry of Agriculture and Forestry (MAF). 2002. "Faster Aerial Spray Operation Planned." *Scoop Independent News*, May 6, 2002.

Ministry of Agriculture and Forestry (MAF). 2003. Environmental Impact Assessment of Aerial Spraying Btk in NZ for Painted Apple Moth.

New Zealand Audit Office. 2003. *Case Study 3 – Response to the Incursion of the Painted Apple Moth.* Wellington, New Zealand.

New Zealand Education Institute (NZEI). 2003. *NZEI West Auckland Aerial Spraying Survey.*

New Zealand Government. 2003a. "Painted Apple Moth eradication." *Scoop Independent News*, April 22, 2003.

New Zealand Government. 2003b. "Painted Apple Moth programme update." *Scoop Independent News*, May 27, 2003.

Noble, Michael A., Peter D. Riben, and Gregory J. Cook. 1992. *Microbiological and Epidemiological Surveillance Programme to Monitor the Health Effects of Foray 48B BTK Spray.* Vancouver, Canada, Ministry of Forests of the Province of British Columbia. https://www.osti.gov/etdeweb/biblio/6657906

Norton, M. L., J. F. Bendell, L. I. Bendell-Young, and C. W. LeBlanc. 2001. "Secondary Effects of the Pesticide *Bacillus thuringiensis kurstaki* on Chicks of Spruce Grouse (*Dendragapus canadensis*)." *Archives of Environmental Contamination and Toxicology* 41: 369–373

Novo Nordisk. 1991, February. *Material Safety Data Sheet for Foray 48B Flowable Concentrate.* Danbury, CT.

Novo Nordisk Enzyme Toxicology Lab. 1990, December 12. *Acute Dermal Toxicity Study in Rabbits with the End Product Foray 48B, Batch BBN 6057.* Danbury, CT.

Office of the Controller and Auditor-General (OAG). 2002. *Report of the Controller and Auditor-General: Management of Biosecurity Risks: Case Studies.* Wellington, NZ: The Audit Office. https://oag.parliament.nz/2002/biosecurity-case-studies/docs/part3.pdf

Office of the Ombudsman. 2007. *Report of the Opinion of Ombudsman Mel Smith on Complaints Arising from Aerial Spraying of the Biological Insecticide Foray 48B on the Population of Parts of Auckland and Hamilton to Destroy Incursions of Painted Apple Moths, and Asian Gypsy Moths, Respectively.* Wellington, NZ: Office of the Ombudsman.

Panckhurst, Paul. 2001. "Mothbeaten." *Metro* October: 57–63.

Patnaik, Pradyot. 2007. *A Comprehensive Guide to the Hazardous Properties of Chemical Substances.* Hoboken, NJ: John Wiley & Sons.

Pearce, Marty, Brian Habbick, Janice Williams, Margaret Eastman, and Maureen Newman. 2002. "The Effects of Aerial Spraying with *Bacillus thuringiensis kurstaki* on children with Asthma." *Canadian Journal of Public Health* 93(1): 21–25.

Pesticide Action Network. 2010. "'Deltamethrin' in PAN pesticide database". Accessed: December 2 2015. http://www.pesticideinfo.org/Detail_Chemical. jsp?Rec_Id=PC33475

Pesticide Action Network Aotearoa. 2002. "MAF admit they don't know if aerial spray is OK." *Scoop Independent News,* October 7, 2002. Accessed April 4, 2022. https://www.scoop.co.nz/stories/AK0210/S00040.htm

Pesticide Action Network of North America (PANNA). 2005. "PANNA CDC Body Burden Study Finds Widespread Pesticide Exposure." http://www.panna. org/legacy/panups/panup_20050722.dv.html

Petrie, Keith, Mark Thomas, and Elizabeth Broadbent. 2003. "Symptom complaints following aerial spraying with biological insecticide Foray 48B." *New Zealand Medical Journal* 116(1170): 1–7. http://www.nzma.org.nz/journal/116-1170/354/.

Pound Sterling Live. n.d. *British Pound/US Dollar Historical Reference Rates from Bank of England for 1975 to 2022.* https://www.poundsterlinglive.com/ bank-of-england-spot/historical-spot-exchange-rates/gbp/GBP-to-USD-1999

Rauh, Virginia, Srikesh Arunajadai, Megan Horton, Frederica Perera, Lori Hoepner, Dana Barr, and Robin Whyatt. 2011. "Seven-Year Neurodevelopmental Scores and Prenatal Exposure to Chlorpyrifos, a Common Agricultural Pesticide." *Environmental Health Perspectives* 119(8): 1196–1201.

Rodenhouse, Nicholas L. and Richard T. Holmes. 1992. "Results of Experimental and Natural Food Reductions for Breeding Black-Throated Blue Warblers." *Ecology* 73(1): 357–372.

Rubio-Infante, Néstor and Leticia Moreno-Fierros. 2016. "An Overview of the Safety and Biological Effects of *Bacillus thuringiensis* Cry Toxins in Mammals." *Journal of Applied Toxicology* 36: 630–648.

Rudzitis, Gundars and Kenton Bird. 2011. "The Myth and Reality of Sustainable New Zealand: Mining in a Pristine Land." *Environment: Science and Policy for Sustainable Development* 53(6): 16–28.

Salama, H. S., A. El-Moursy, F. N. Zaki, R. Aboul-Ela, and A. Abdel-Razek. 1991. "Parasites and Predators of the Meal Moth *Plodia interpunctella* Hbn. as Affected by *Bacillus thuringiensis* Berl." *Journal of Applied Entomology* 112: 244–253.

Sanahuja, Georgina, Raviraj Banakar, Richard M. Twyman, Teresa Capell, and Paul Christou. 2011. "*Bacillus thuringiensis*: A Century of Research, Development and Commercial Applications." *Plant Biotechnology Journal* 9(3): 283–300.

Saskatchewan Environmental Society (SES). 2003. *The Case against Overhead Pesticide Spraying in the Townsite of Waskesiu Lake, Prince Albert National Park of Canada.* http://gaiavisions.org/GypsyMoths/panp-spraying-1.pdf

Savonen, Carol. 1994. "Btk spraying for forest pest kills many other species." *OSU News.* Corvallis, OR: Oregon State 'University' Agricultural Communications.

Sharma, Poonam, Rambir Singh, and Mysra Jan. 2014. "Dose-Dependent Effect of Deltamethrin in Testis, Liver, and Kidney of Wistar Rats." *Toxicology International* 21(2): 131–139.

Siegel, Joel P. 2001. "The Mammalian Safety of *Bacillus thuringiensis*-based Insecticides." *Journal of Invertebrate Pathology* 77(1): 13–21.

Siegel, Joel P., John A. Shadduck, and James Szabo. 1987. "Safety of the Entomopathogen *Bacillus thuringiensis var. israelensis* for Mammals." *Journal of Economic Entomolgy* 80(4): 717–723.

Smith, Barbara Ellen. 1981. "Black Lung: The Social Production of Disease." *International Journal of Health Services* 11(3): 343–359.

Solomon, Gina. 2000. *Pesticides and Human Health: A Resource for Health Care Professionals* [online]. Berkeley, CA: Physicians for Social Responsibility. Retrieved December 10, 2015. http://www.psr-la.org/files/pesticides_and_human_health.pdf

Suckling, David, John Charles, Malcolm Kay, John Kean, Graham Burnip, Asha Chhagan, Alasdair Noble, and Anne Barrington. 2014. "Host Range Testing for Risk Assessment of a Sexually Dimorphic Polyphagous Invader, Painted Apple Moth." *Agricultural and Forest Entomology* 16: 1–13.

Surgeoner, Gordon A. and Martha Judit Farkas. 1990. "Review of *Bacillus thuringiensis* var. kurstaki (BTK) for Use in Forest Pest Management Programs in Ontario-with Special Emphasis on the Aquatic Environment."

Swadener, Carrie. 1994. "*Bacillus thuringiensis* (BT) Insecticide Fact Sheet." *Journal of Pesticide Reform* 14(3): 13–20.

Tayabali, Azam and Verner Seligy. 2000. "Human Cell Exposure Assays of *Bacillus thuringiensis* Commercial Insecticides: Production of *Bacillus cereus*-like Cytolytic Effects from Outgrowth of Spores." *Environmental Health Perspectives* 108(10): 919–930.

Tos-Luty, Sabina, Agnieszka Haratym-Maj, Jadwiga Latuszynska, Daniela Obuchowska-Przebirowska, and Malgorzata Tokarska-Rodak. 2001. "Oral Toxicity of Deltamethrin and Fenvalerate in Swiss Mice." *Annals of Agricultural and Environmental Medicine* 8: 245–254.

Tsai, S. F., L. J. Wang, and S. C. Wang. 1997. "Clearance and Effects of Intratracheal Instillation to Spores of *Bacillus thuringiensis* or *Metarhizium anisopliae* in Rats." *Journal-Chinese Society of Veterinary Science* 23: 515–522.

Tuzmen, Nalan, Nulgün Candan, Eren Kaya, and Nazan Demiryas. 2008. "Biochemical Effects of Chlorpyrifos and Deltamethrin on Altered Antioxidative Defense Mechanisms and Lipid Peroxidation in Rat Liver." *Cell Biochemistry and Function* 26(1): 119–124.

Tyson, Janet. 2009. "The Painted Apple Moth Eradication Programme (Vignette Version: A)." *The Case Program, The Australian and New Zealand School of Government.* www.anzsog.edu.au

U.S. Environmental Protection Agency. 1998. *Reregistration Eligibility Decision (RED) Bacillus thuringiensis.* March. https://nepis.epa.gov/Exe/ZyPDF.cgi/2000O6G.PDF?Dockey=2000O6G.PDF

Upton, Roy and Lynette Caspar. 2008. "*Bacillus thuringiensis* – Safety Review." Citizens for Health. Accessed January 23, 2016. http://www.lbamspray.com/Reports/BacillusthuringiensisSafetyReview031208.pdf

Valadares de Amorim, Giovana, Beatrixe Whittome, Benjamin Shore, and David Levin. 2001. "Identification of *Bacillus thuriengiensis* subsp. *kurstaki* Strain HD1-Like Bacteria from Environmental and Human Sample after Aerial Spraying of Victoria, BC, Canada with Foray 48B." *Applied and Environmental Microbiology* 67(3): 1035–1043.

Van Cuyk, Sheila, Alina Deshpande, Attelia Hollander, Nathan Duval, Lawrence Ticknor, Julie Layshock, LaVerne Gallegos-Graves, and Kristin Omberg. 2011. "Persistence of *Bacillus thuringiensis* subsp. *kurstaki* in Urban Environments following Spraying." *Applied and Environmental Microbiology* 77(22): 7954–7961.

Velsicol Chemical Corporation. 1981. *Four Week Subacute Inhalation Toxicity Study of Benzoic Acid in Rats*. Report prepared by International Research and Development Corporation, Mattawan, MI, for Velsicol Chemical Corporation, Chicago, IL (FYI-OTS-1281-0147).

Vettori, Cristina, Donatella Paffetti, Deepak Saxena, Guenther Stotzky, and Raffaello Giannini. 2003. "Persistence of Toxins and Cells of *Bacillus thuringiensis* subsp. *kurstaki* Introduced in Sprays to Sardinia Soils." *Soil, Biology & Biochemistry* 35: 1635–1642.

Vizcaya, David, Maria Mirabelli, Josep-Maria Antó, Ramon Orriols, Felip Burgos, Lourdes Arjona, and Jan-Paul Zock. 2011. "A Workforce-Based Study of Occupational Exposures and Asthma Symptoms in Cleaning Works." *Occupational and Environmental Medicine* 68(12): 914–919.

Wang, Zhi-zhi, Yin-quan Liu, Min Shi, Jian-hua Huang, and Xue-xin Chen. 2019. "Parasitoid Wasps as Effective Biological Control Agents." *Journal of Integrative Agriculture* 18(4): 705–715.

Washington State Department of Health. 1993, March. *Report of Health Surveillance Activities: Asian Gypsy Moth Control Program*. Olympia, WA.

Washington State Department of Health. 2001. *Report of Health Surveillance Activities: Aerial Spraying for Asian Gypsy Moth – May 2000 Seattle, WA*. https://www.doh.wa.gov/Portals/1/Documents/Pubs/334-292.pdf

Watts, Meriel. 2003. *Painted Apple Moth Eradication Programme: Health Risk and Effects*. https://www.semanticscholar.org/paper/PAINTED-APPLE-MOTH-ERADICATION-PROGRAMME-%3A-HEALTH-Watts /758c4f98b73b1e6190d09facf8639763ee68a805Zargar

West, Andrew W. and H. Denis Burges. 1985. "Persistence of *Bacillus thuringiensis* and *Bacillus cereus* in Soil Supplemented with Grass or Manure." *Plant and Soil* 83: 389–398.

West Aucklanders Against Aerial Spraying (WASP). 2002a. "Painted Apple Moth Foray 48b aerial spraying." *Scoop Independent News,* August 24, 2002. Accessed April 4, 2022. http://www.scoop.co.nz/stories/PO0208/S00094.htm

West Aucklanders Against Aerial Spraying (WASP). 2002b. "Painted Apple Moth Foray 48b aerial spraying." *Scoop Independent News,* December 4, 2002. Accessed May 15, 2022. https://www.scoop.co.nz/stories/AK0212/S00023.htm

Wibbertmann, A., J. Kielhorn, G. Koennecker, I. Mangelsdorf, and C. Melber. 2000. *Concise International Chemical Assessment Document No. 26: Benzoic Acid and Sodium Benzoate*. Geneva: International Programme on Chemical Safety INCHEM. https://inchem.org/documents/cicads/cicads/cicad26.htm

Young, K. 1994. "1993–1994 Urban aerial spraying of Foray 48B: Site Inspection and observations in Victoria/Saanich, BC, Canada." Report for the Ecological Health Alliance. Revised April 1999. Accessed 6/7/2015. http://www.ehabc.org/pdfs/ehabc_report_aerial_btk_rev1999.pdf

2 The Social Production of a Foreign Species Incursion

In recent decades invasive species have become a growing concern, as there has been a surge of unwanted foreign species migrating into new locales. Notable examples include the spread of zebra mussels into the Great Lakes, the Mediterranean fruit fly in California and New Zealand, the Gypsy Moth in the Pacific Northwest and New Zealand, the Light Brown Apple Moth in California, the Argentine ant in numerous countries (including the United States, South Africa, Australia, and New Zealand), and the Asian tiger mosquito in numerous countries (including Brazil, Southern Europe, South Africa, Australia, and the Southeastern United States) (Green 2000; Gutierrez & Ponti 2011; Spiegelman 2010; Parliamentary Commissioner for the Environment (PCE) 2000). These invasions are concerning because of the potential threat they represent to agriculture, native ecology, economy, and, in some cases, public health (Jay et al. 2003; Pimental 2002; Schmitz and Simberloff 1997).

Although the painted apple moth's incursion was the environmental problem being addressed by the 2002–2004 aerial pesticide spraying campaign, a deep understanding of pesticide use requires considering how the environmental problem being addressed (i.e. the PAM incursion) was itself socially produced. As McCarthy Auriffeille and King (2014) emphasize, environmental problems are not random occurrences but rather are significantly mediated by human actions, decision-making, and social systems. This is true whether we are considering flooding, wildfires, a foreign species incursion, or other environmental problems.

As this pertains to foreign species incursions, biosecurity researchers have emphasized it is not individual species that are invasive but rather the "socio-biological networks" that enable such invasions to take place (Jay & Morad 2006; Robbins 2004). Moreover, they emphasize it would be more useful to study human ecologies of species invasion than to focus solely on individual species. That is to say, instead of focusing solely on eliminating species that make their way into new locales, we should direct our attention to the suite of human relations, activities, systems, and policies that create favorable conditions for invasions to occur. Regarding the PAM incursion, it was mediated by numerous human factors and this chapter will discuss six:

DOI: 10.4324/9780429426414-3

the globalization of trade, technological developments, inadequate biosecurity systems, increasing human migration, a political economy that prioritizes economic growth, and an ineffective initial response to the moth's discovery.

The Globalization of Trade

An important trend in the second half of the twentieth century was the increased globalization of trade, which increased, in three ways, the risk of foreign species traveling to new countries.

First, globalization increased the number of countries and ports participating in the global shipping network, which added new links to the shipping routes, with each new port of call representing a new pathway through which unwanted species could be transported to New Zealand (Green 2000).

It is important to note that some pathways are riskier than others. One mediating factor is that some countries have a higher number of exotic species, as is the case with Australia and countries in Asia and Latin America. Another mediating factor is cultural differences in biosecurity concerns and approaches, which can lead biosecurity protocols in some countries to ignore issues of concern to other countries (Jay et al. 2003). For instance, while particular insects might be of significant concern in some countries they can be of much less concern in other countries, which can lead their officials to be unrigorous in their application of the rules. A third factor mediating a country's riskiness is the underdevelopment and/or underfunding of its biosecurity systems. This is often the case with developing countries, which are often still reeling from the vestigial economic effects of colonialism and oppressive post-colonial institutions (such as International Monetary Fund loan restructuring programs) that prevent them from adequately funding their basic infrastructure (such as education and healthcare), let alone biosecurity programs (Davis 2010; Diaz 2010; Perkins 2016; Sachs 2005).

Second, globalization increased the volume of global trade (including exports and imports) since World War II (WWII). Increased trade is viewed as the main driver of alien species transmissions, and such transmissions tend to occur either through the transfer of ballast water, species establishing a foothold on ship hulls, or goods shipped in containers (Bertelsmeier 2021; Jenkins 1996; Mack et al. 2000; Myerson & Mooney 2007; Perrings et al. 2005). The latter is the principal transmission method for insects and, given this book's focus on the painted apple moth, is the pathway I focus on.

A robust indicator of growing global trade is that it has represented a growing share of global Gross Domestic Product (GDP) since WWII: where global trade represented 10% of global GDP in 1945, the figure grew to 47% by 2000 (Ortiz-Ospina & Beltekian 2018a). As it pertains to New Zealand, in the decade prior to PAM's arrival, global trade's percentage of GDP rose from 48% in 1989 to 68% in 2000 (Ortiz-Ospina & Beltekian 2018b; Trading Economics

2000a). Although New Zealand is known for *exporting* lamb, dairy, fruits, vegetables, and wood products, the country's *imports* also grew substantially in the two decades preceding the PAM incursion, increasing from NZD $250 million in 1980 to over NZD $2.5 billion in 2000 (Trading Economics 2020b). Another indicator of growing imports is that in the five years leading up to the PAM incursion the number of sea containers entering New Zealand increased by 114%, growing from approximately 175,000 in 1994 to over 375,000 in 1999 (Budd and Arts 2000; Jay et al. 2003). Moreover, the gross weight of overseas cargo unloaded doubled between 1991 and 2000 (PCE 2000).

Shipping goods from one country to another is a significant biosecurity risk, as species from one locale can stowaway in the shipped goods, packing materials, or containers. Examples include insects that reside on wood products, mosquito larvae existing in water found in second-hand tires, moths and other insect colonies harboring in imported cars, snakes that have infiltrated shipping containers, as well as fungi and nematodes, which can be found in soil stuck to the bottom of containers (Green 2000; Jenkins 1996; PCE 2000).

Researchers have uncovered that it is through imports that most non-indigenous species arrive (Convention on Biological Diversity 2002; Daehler and Carino 1999; Jenkins 1996), with the United States Office of Technology (1993) reporting that, between 1980 and 1993, 81% of harmful new exotics were unintentionally brought in through trade. Moreover, an increase in global trade has been found to raise the risk of propagation to new countries, as it increases both the number of introduction events and the number of individuals from a species that are introduced to new locales (Jenkins 1996, 1999; Vilà & Pujadas 2001). This would be particularly true in countries with underfunded biosecurity systems, which were the vast majority of both industrialized and developing countries at that time.

Another problem created by globalization is that it expanded the number of products destined for New Zealand, which substantially increased the number of products that provide a pathway for transmitting unwanted organisms (PCE 2000). A particularly problematic product is used cars, which are particularly susceptible for introducing new species, as they can harbor insects and/or insect egg masses, such as Asian gypsy moths, nun moths (*Lymantria monacha*), and white-spotted tussock moths (*Orgyia thyellina*), each of which have been detected on used car imports (Armstrong et al. 2003; Dann 2002; PCE 2000). Used cars are particularly germane to the New Zealand case because they make up a significant percentage of imports. Specifically, in 1997 they represented the second highest import by value (12.7%) (PCE 2000). Moreover, in the years preceding the PAM incursion the importation of cars grew significantly, growing from 50,000 in 1994 to 120,000 in 1999 (Dann 2002; PCE 2000). Beyond being a potential source of new transmissions, the increased importation of used cars will have siphoned biosecurity resources, which will be unavailable for addressing other biosecurity concerns.

Technological Developments

During the 1990s the risk of foreign species incursions was further increased by technological developments, including improvements in transportation (air transport and faster ships), which decreased transit times and increased the chances of foreign species surviving the trip (Myerson & Mooney 2007; PCE 2000).

Another major technological development was the shipping industry's adoption of containerization, which revolutionized the industry because it allowed shippers to have containers off-loaded onto rail or trucks and transported directly to buyers, thereby increasing the speed at which goods can be transported from source to market. Between 1980 and 2000 containers arrived in New Zealand at a faster rate than the substantial rate at which global trade was expanding (Green 2000).

A problem with containers is that unwanted species can infiltrate into them and establish a foothold. Species that have been found in New Zealand-bound containers include snakes, scorpions, spiders, slugs, moths, other insects, as well as weeds (Green 2000; PCE 2000). Compounding this problem, containers are very difficult to properly inspect when fully packed. Moreover, the ease with which containers can be off-loaded and transported to final destinations means containers, and the species they contain can travel deeply into a country before the containers ever get opened, let alone inspected, thereby giving an invasive species an opportunity to spread extensively before being discovered. This problem is further exacerbated in countries that only inspect a fraction of imported containers, as was the case in New Zealand during the 1990s, when only a quarter of containers were being inspected (Green Party 2003; Office of the Controller and Auditor-General (OAG) 2002).

Besides the invasive species that might be hitchhiking inside the containers, foreign organisms can be introduced via the soil stuck to the bottom of the containers, such as fungi and nematodes, which are considered a risk to New Zealand Forestry (PCE 2000). A 1997–1998 study of risks to forestry found that 23% of 3,681 containers carried quarantinable contaminants and that the rate was higher for certain countries/regions, such as Australia (28.3%), Southeast Asia (33.2%), and South Africa (50%) (Green 2000; OAG 2002).

This problem is exacerbated by three issues associated with containers: (1) they can be unloaded and railed to a distant location before being unpacked and inspected; (2) containers are rarely cleaned between shipments, thereby making them ideal vectors for transporting species between locations; (3) containers can be stored in one place for lengthy periods, thereby giving unwanted organisms an opportunity to establish a foothold before being transported to the next location (Green 2000).

Inadequate Biosecurity Protection

Another factor that mediates species migration is the biosecurity response apparatus in that locale. In New Zealand's case, the Parliamentary Commissioner for the Environment (2000) considered the country's biosecurity forces to be limited in their ability to effectively monitor the growth of incoming traffic, detect possible incursions, and respond quickly and effectively. Part of the problem was the proliferation of entry points for goods and people, which made it difficult for biosecurity agencies to keep out unwanted foreign species (*Ibid.*). This problem is particularly salient to an island nation like New Zealand, which has a lot of private boat traffic and 16 provincial ports (Green 2000).

Also contributing to the problem was the government's unwillingness to allocate sufficient funds to protect against the biosecurity risks created by increased traffic. During the 1990s the rate of imports grew faster than the rate of spending on surveillance, monitoring, and quarantine operations (Green 2000). This resulted in insufficient staffing, technically limited border detection systems, and inadequate inspection of incoming containers (Office of the Controller and Auditor-General (OAG) 2002; PCE 2000). For instance, even though MAF acknowledged that 39% of containers were contaminated, they only inspected 24% of them (Green Party 2002, 2003).

The failure to properly fund biosecurity suggests there was a significant gap existing between New Zealand's actions and symbolic aspirations. While the country set a world precedent by passing the 1993 Biosecurity Act, the government had, by 2001, proven unwilling to pay the cost required to adequately protect the country from the biosecurity risks introduced by expanding global trade. Moreover, the funding that was being provided was geared disproportionately towards protecting the agricultural or forestry industries, with little funding going to other biosecurity concerns. This is reflected in the fact that, as of 2000, the Ministry of Agriculture and Forestry (MAF) was receiving 95% of biosecurity funding, while the Ministry of Conservation only received 5% (PCE 2000).

Globalization: Increases in International Travel

Another pertinent aspect of globalization has been the surge of human movement into New Zealand, either through migration, tourism, or New Zealanders returning from tourism in other countries. Such movements grew sizably in the late twentieth century. For example, the total number of human trips (including both international tourists and returning residents) to New Zealand more than tripled between 1982 and 1999 (the year of PAM's incursion), increasing from 914,257 to 2,876,610, and continued to grow in subsequent years, reaching 7,100,373 in 2019 (Stats NZ 2021a). There was a similar growth in international tourists, with numbers rising from 465,163 in 1980 to 1,607,478 by 1999, and to 3,888,473 by 2019 (Stats NZ 2021b).

An increase in human movement into countries is a biosecurity concern because of the foreign species they can bring through their personal belongings. While the most common examples are of people inadvertently bringing fresh fruit into the country, another concern is the inadvertent importing of foreign species in luggage. For example, camping and other sports equipment can contain live insects, plant seeds, and pathogenic fungi (PCE 2000).

Underscoring this point, in December 1981 researchers found that 13% of tents belonging to air passengers entering the country contained live insects (Gadgil & Flint 1983). This concern is particularly relevant to New Zealand, given the number of tourists who come for outdoor activities and who bring their own sporting equipment.

Although inadvertence is part of the problem, another is obliviousness, as many travelers come from countries where the concept of "invasive species" is mostly unknown and are unlikely to be aware of New Zealand's biosecurity requirements. In turn, this inclines them to bring back problematic species without thinking twice about the potential biosecurity risk it creates for New Zealand. A common example is passengers trying to bring back fresh fruit and/or honey from their foreign travels, without realizing these are prohibited items. A more exotic example is the passenger who was unaware of New Zealand's biosecurity requirements and brought back a live giant African snail for a gourmet meal (Green 2000).

Another part of the problem is carelessness, as some are aware of the biosecurity restrictions but disregard them. An example is travelers knowing about New Zealand's prohibition of introducing fruit into the country but continuing to do so. A more extreme example would be smugglers who bring in banned products knowing fully well they are banned substances. An example is beekeepers illegally importing queen bees, which can carry a variety of pathogens. It is believed the establishment of the varroa mite in New Zealand, in 2000, was due to such illegal activity (Iwasaki et al. 2015). Given the varroa mites' economic impact on the country's beekeeping industry, one could argue that the smuggler was probably unaware of the potential environmental and economic impacts associated with importing queen bees. Then again, some smugglers may be keenly aware of potential environmental impacts, as was the case with those who imported rabbit hemorrhagic disease, which was then released in the country with the hope that it would decimate the country's burgeoning rabbit population (*Ibid.*).

Although the problems associated with growing human travel did not directly contribute to painted apple moth's arrival onto New Zealand shores, the increase in international travelers and the biosecurity risks they produce put a growing strain on the limited biosecurity resources. In turn, this is likely to have siphoned off resources needed to adequately inspect shipped goods.

A Growth-Oriented Political Economy

A comprehensive analysis should also consider the role of the political economy. Previous sociologists have elucidated that a capitalist political economy encourages participants to constantly strive to increase economic growth, which is a reason countries tend to fixate on maximizing Growth Domestic Product (GDP), often at the expense of health and environmental considerations

(Carolan 2020; Schnaiberg & Gould 1994). The focus on economic growth is reflected in the surge of global trade that has occurred since WWII, both in New Zealand and elsewhere, which has increased GDP. It is also reflected in the growth of New Zealand's tourism industry, which significantly increased the number of international visitors and the country's GDP.

Another characteristic of growth-oriented economic systems is the tendency for governments to subsidize industries by assuming the costs of harmful industry practices. This includes practices that are harmful to workers, consumers, and those in surrounding communities and/or the environment. An example is government agencies that remediate water pollution caused by paper mills, which dump waste products, such as mercury, into rivers. Another example is having the healthcare system, and the taxpayers who fund it, pick up the cost of treating workers and/or community members harmed by industrial pollution. In both cases, harm is caused by a corporation continuing to use harmful practices, but whose costs are successfully externalized to the government. To the extent governments pick up such costs, they are subsidizing industry and are incentivizing it to continue using harmful practices. Moreover, it is disincentivizing the pursuit of less harmful practices, as developing such practices would require investing money that would place a company at an economic disadvantage vis-a-vis its adversaries. Conversely, if industries had to pay the full cost of fixing the damage caused by their harmful business practices, it would completely alter the playing field and would incentivize corporations to develop less harmful practices.

As it pertains to the shipping industry, when a foreign species gains a foothold in a new country, it is not the shipping company that is tasked with paying for the containment and eradication of that species. Rather, that burden falls on government agencies and the tax-paying public, which encourages companies to continue using risky shipping practices. However, if the situation was reversed so that shippers had to pay the damage caused by outbreaks of foreign organisms, the shipping industry would be incentivized to develop practices that minimize the transportation of unwanted organisms into new countries (Perrings et al. 2005).

Another important aspect of political economy is the organizations that perpetuate the system's tendencies. One such organization is the World Trade Organization (WTO), which seeks to maximize trade by eliminating both national trade restrictions designed to protect the country's environment and any environmental regulations that could inhibit trade (PCE 2000). For example, if country A feels that their ability to trade with country B is inhibited by the latter's environmental regulations, country A can take the other country to WTO court. Thus, trade agreements discourage countries from passing or enforcing environmental regulations, which is particularly problematic for countries with acute biosecurity concerns like New Zealand (PCE 2000).

To be fair, WTO agreements are not necessarily opposed to environmental concerns, and the WTO has directed some complaints to be resolved under Multilateral Environmental Agreements (MEAs) that both parties have signed on to. For instance, the WTO argues:

> While the WTO members have the right to bring disputes to the WTO dispute settlement mechanism, if a dispute arises between WTO members... over the use of trade measures they are applying to themselves pursuant to the MEA (Multilateral Environmental Agreement), they should consider trying to resolve it through the dispute settlement mechanisms available under the MEA.
>
> (WTO 1996, as cited by PCE (2000), p. 73)

However, as the New Zealand Parliamentary Commissioner for the Environment emphasized:

> While this suggests that the MEA process is preferable in some cases, there is no guarantee that a party will not invoke the WTO dispute resolution process where that party sees its own trade interests being affected.
>
> (PCE 2000, p. 73)

Additionally, if the complaint does go to WTO dispute resolution, the WTO favors the use of "science" to adjudicate the decision, where the party seeking to restrict trade will have the burden of proving their proposed restriction is scientifically justifiable, as determined by three evaluative institutions recognized by the WTO (*Ibid.*). This is problematic on three levels. Not only is the burden placed on those trying to protect their ecosystems, but also the reliance on science invariably means there needs to be proof of damage after the fact, which is the opposite of the precautionary principle most MEAs are based on. And third, the WTO is the arbiter of who is allowed to evaluate the "science" in question, which can lead to subjective and opaque decisions. A far more effective arrangement would be to have the science evaluated by a blue-ribbon panel of recognized scientific experts who have no industry ties.

As a signatory to the 1994 Uruguay WTO agreement negotiations, New Zealand had signed on to the process of expanding trade and would have been hard-pressed in the late 1990s to introduce environment-protecting legislation that would have inhibited the growth of trade imports.

The Ineffective Biosecurity Response

Besides understanding the contextual factors that contribute to producing a foreign species incursion, we also need to consider the effectiveness of a government's response to such an incursion.

As covered in the previous chapter, after the moth's original detection, in April 1999, MAF pursued a program that consisted of visually searching for moths on plant and trees, destroying host trees where painted apple moths were found, and spraying chlorpyrifos and deltamethrin insecticides (Goven et al. 2007). However, this operation was ineffective, as underscored by the fact the moth had not been eradicated two years after its initial incursion.

MAF's failure to not only eradicate the moth but to contain its spread led it to fall under increasing pressure, particularly from the forestry industry. In response to industry pressure, MAF commissioned an independent review, which concluded that while personal conflicts had compromised the eradication operation, the overall eradication strategy was appropriate (Goven et al. 2007). However, the Office of the Controller and Auditor-General (OAG) was far more critical in its assessment, arguing MAF's "response contained flaws and errors of judgement made by senior MAF staff" (OAG 2002, p. 65). Specifically, the OAG critiqued five aspects of MAF's response: (1) it dedicated insufficient resources for the response; (2) it rejected offers of assistance from experts because of personal disputes; (3) it used poor documentation standards; (4) MAF senior management failed to exercise adequate management control and oversight over the response to PAM; and (5) the agency failed to consult in a timely manner with the community about its proposed actions to eradicate PAM (OAG 2002).

What we now know about infestations is that it is important to address the problem early and forcefully. So, the unwillingness to allocate sufficient funding seems, in hindsight, to have been a particularly poor decision. Compounding that problem was that the PAM operation director (i.e. Ruth Frampton) refused to use experts who offered their help at the very beginning of the infestation and whose expertise would have accelerated the eradication operation (OAG 2002).

Regarding the latter, there were at least two instances where MAF rejected expert assistance. First, they declined the assistance of John Clearwater and Professor Gerhard Gries to develop biological controls for the painted apple moth. This was a fateful decision as Clearwater and Gries were considered world experts on the control of lymantriid moths, having developed the mating disruption technology that was instrumental to eradicating the white-spotted tussock moth that plagued Auckland in 1996 and 1997 (Goven et al. 2007; OAG 2002). The technology consisted of developing pheromones to lure males to sterile females and would have considerably stymied the spread of the painted apple moth. In April 1999, after hearing about the PAM's discovery in Auckland, John Clearwater phoned and mailed the MAF director to offer their services for free (OAG 2002). However, the director declined to even respond to their offers and instead gave the work to another laboratory, which, presumably, she had social relationships with and which failed to produce the pheromone. Even more stunning is that MAF refused to give Clearwater and Gries the work even after the independent reviewers strongly recommended MAF use every means at their disposal to secure the services of these researchers (*Ibid.*).

The Controller and Auditor-General (2002) took MAF, and in particular the Director of Forest Biosecurity (Ruth Frampton), to task not only for failing to accept the services of world experts who had offered their services for free but also for showing poor form in failing to even respond to the experts' offer. While the Auditor-General attributed the problems to the poor working relationships between the Director of Forest Biosecurity and key players in the sector, he emphasized that MAF's "priority in responding to exotic pests should be to ensure that the chances of a successful response are maximised" (OAG 2002, p. 68). Moreover, he stated that if Clearwater's offer to help had been accepted "there is a strong likelihood that the response to PAM would have been different" (Controller and Auditor 2002, p. 69).

The second expert offer MAF rejected was from Forest Research labs, which offered, at the outset of the PAM incursion, to breed a colony of painted apple moths, at no expense to MAF, which could be used for feeding trials and for developing mating disruption technology. As well, they offered to host feeding trials at its new quarantine facility, again, at no expense to MAF. However, the agency declined. Four months later, in August 1999, MAF contracted a different team of scientists to rear the moths and develop a pheromone. This new team experienced delays in developing a breeding colony, which stalled the development of a pheromone and undermined the operation's chances of success (Goven et al. 2007). Consequently, these blunders allowed PAM to spread well beyond the original infestation sites. As was the case with the Clearwater and Gries offer, in May 2001 the independent reviewers recommended that MAF contract with Forest Research labs to establish a breeding colony (Goven et al. 2007). However, by that point MAF was already proceeding with plans for an aerial pesticide spraying operation (*Ibid.*).

Summary

The main purpose of this chapter has been to trace the socio-biological configuration within which the PAM incursion occurred, with a particular focus on identifying the social factors that increased the likelihood an invasive species would establish and spread in New Zealand. Towards that end, I discussed the role of increased global trade, technological developments in shipping, inadequate biosecurity resourcing, growing levels of international travel (by both New Zealanders and international visitors), a political economy geared towards maximizing economic growth, and the biosecurity agency's failure to contain the moth's spread at the outset.

Beyond the PAM case, this analysis has implications for understanding other cases where societies experience the arrival of a foreign species. First, the analysis underscores that the arrival of a foreign species is not a random occurrence but rather is mediated by economic, technological, political, and social configurations. In turn, this suggests that analyses of aerial pesticide spraying operations should not only study the deployment of pesticide spraying but should also analyze the social factors that enabled the incursion to happen in the first place.

Many experts maintain it is cheaper to prevent invasive species from arriving into the country than it is to manage the problem after the fact (Mack et al. 2000). However, countries perpetuate systems, policies, and practices that reproduce the problem that is to be avoided. If the intent is to prevent such incursions from happening, then it behooves countries to identify, understand, and effectively address the social factors that continue to reproduce the problem, not the least of which are risky shipping industry practices, the state's unwillingness to increase shipping regulations, the state's unwillingness to properly fund biosecurity monitoring, and WTO policies that are at odds with the precautionary principle.

Now that we better understand the social factors contributing to PAM's arrival, the next step is to understand why New Zealand targeted it for eradication, which was not a given. After all, since the start of colonization (i.e. 1840) it is estimated 19,000+ foreign species have been introduced into New Zealand, the vast majority of which have not been targeted for eradication. Moreover, in its native Australia the painted apple moth is only considered a *minor* garden pest and is not targeted for control, much less eradication. The next chapter unpacks New Zealand's intolerance towards this particular species, illuminating the contextual factors that contributed to its response.

References

Armstrong, Karen F., P. McHugh, W. Chinni, E. Ruth Frampton, and Patrick J. Walsh. 2003. "Tussock Moth Species Arriving on Imported Used Vehicles Determined by DNA Analysis." *New Zealand Plant Protection* 56: 16–20.

Bertelsmeier, Cleo. 2021. "Globalization and the Anthropogenic Spread of Invasive Social Insects." *Current Opinion in Insect Science* 46: 16–23.

Budd, K. and A.-M Arts. 2000. *Review of Biosecurity Influences of the Last Decade.* Unpublished report prepared for the Parliamentary Commissioner for the Environment, Wellington, New Zealand.

Carolan, Michael. 2020. *Society and the Environment: Pragmatic Solutions to Ecological Issues, 3rd Edition.* New York: Routledge Press.

Controller and Auditor-General (OAG). 2002. "Report of the Controller and Auditor General." Management of Biosecurity Risks: Case Studies. November 2002.

Convention on Biological Diversity (CBD). 2002. *Assessing the Impact of Trade Liberalization on the Conservation and Sustainable Use of Agricultural Biological Diversity.* CBD: The Hague.

Daehler, Curtis C. and Debbie A. Carino. 1999. "Threats of Invasive Plants to the Conservation of Biodiversity." In *Biodiversity and Allelopathy: From Organisms to Ecosystems in the Pacific,* edited by C. Chang-Hung, G.R. Waller, and C. Reinhardt, 21–27. Taipei: Academia Sinica.

Dalsager, Louise, Bettina Fage-Larsen, Niels Bilenberg, Tina Kold Jensena, Flemming Nielsen, Henriette Boye Kyhl, Philippe Grandjean, and Helle Raun Andersen. 2019. "Maternal Urinary Concentrations of Pyrethroid and Chlorpyrifos

Metabolites and Attention Deficit Hyperactivity Disorder (ADHD) Symptoms in 2–4-Year Old Children from the Odense Child Cohort." *Environmental Research* 176: Article 108533.

Dann, Christine. 2002. "Losing Ground? Environmental Problems and Prospects at the Beginning of the Twenty-First Century." In *Environmental Histories of New Zealand*, edited by Eric Pawson and Tom Brooking, 275–287. Oxford University Press.

Davis, Mike. 2010. *Late Victorian Holocausts: El Niño Famines and the Making of the Third World*. London, UK: Verso Press.

Diaz, Philippe. 2010. *The End of Poverty? Think Again*. Canoga Park, CA: Cinema Libre Studio.

Gadgil, Peter D. and T. N. Flint. 1983. "Assessment of the Risk of Introduction of Exotic Forest Insects and Diseases in Imported Tents." *New Zealand Journal of Forestry* 28: 58–67.

Goven, Joanna, Tom Kerns, Romeo Quijano, and Dell Wihongi. 2007. *Report of the March 2006 People's Inquiry Into the Impacts and Effects of Aerial Spraying Pesticide over Urban Areas of Auckland, for the People's Inquiry*.

Green Party. 2002. "Too much wiggle-room at the biosecurity border." *Scoop Independent News*, November 28, 2002. http://www.scoop.co.nz/stories/PA0211/S00643.htm

Green Party. 2003. "Simple solution to shipping incursions." *Scoop Parliament*, February 26. https://www.scoop.co.nz/stories/PA0302/S00481.htm

Green, Wren. 2000. *Biosecurity Threats to Indigenous Biodiversity in New Zealand – An Analysis of Key Issues and Future Options*. A Background report prepared for the Parliamentary Commissioner for the Environment, Wellington.

Gutierrez, Andrew Paul and Luigi Ponti. 2011. "Assessing the Invasive Potential of the Mediterranean Fruit Fly in California and Italy." *Biological Invasions* 13: 2661–2676.

Iwasaki, Jay M., Barbara I. P. Barratt, Janice M. Lord, Alison R. Mercer, and Katharine J. M. Dickinson. 2015. "The New Zealand Experience of Varroa Invasion Highlights Research Opportunities for Australia." *Ambio* 44(7): 694–704.

Jay, Mairi and Munir Morad. 2006. "The Socioeconomic Dimensions of Biosecurity: The New Zealand Experience." *International Journal of Environmental Studies* 63(3): 293–302.

Jay, Mairi, Munir Morad, and Angela Bell. 2003. "Biosecurity: A Policy Dilemma for New Zealand." *Land Use Policy* 20(2): 121–129.

Jenkins, Peter T. 1996. "Free Trade and Exotic Species Introductions." *Conservation Biology* 10: 300–302.

Jenkins, Peter T. 1999. "Trade and Exotic Species Introductions." In *Invasive Species and Biodiversity Management*, edited by O. T. Sandlund, P. Schei, 229–235. Kluwer Academic Publishers, The Netherlands.

Lu, Qirong, Yaqi Sun, Irma Ares, Arturo Anadón, Marta Martínez, María-Rosa Martínez-Larrañaga, Zonghui Yuan, Xu Wang, and María-Aránzazu Martínez. 2019. "Deltamethrin Toxicity: A Review of Oxidative Stress and Metabolism." *Environmental Research* 170: 260–281.

Mack, Richard N., Daniel Simberloff, W. Mark Lonsdale, Harry Evans, Michael Clout, and Fakhri A. Bazzaz. 2000. "Biotic Invasions: Causes, Epidemiology, Global Consequences, and Control." *Ecological Applications* 10(3): 689–710.

McCarthy Auriffeille, Deborah and Leslie King. 2014. "Introduction: Environmental Problems Require Social Solutions." In *Environmental Sociology: From Analysis to Action, 3rd Edition*, edited by Leslie King and Deborah Auriffeille McCarthy, 1–23. Lanham, MD: Rowman & Littlefield Publishers.

Myerson, Laura A. and Harold A. Mooney. 2007. "Invasive Alien Species in an Era of Globalization." *Frontiers in Ecology and the Environment* 5(4): 199–208.

Office of Technology Assessment, U.S. Congress. 1993. *Harmful Non-Indigenous Species in the United States*. Washington, DC: U.S. Government Printing Office.

Office of the Controller and Auditor-General. 2002. *Report of the Controller and Auditor-General: Management of Biosecurity Risks: Case Studies*. Wellington, NZ: The Audit Office. https://oag.parliament.nz/2002/biosecurity-case-studies/docs/part3.pdf

Ortiz-Ospina, Esteban and Diana Beltekian. 2018a. "Trade and Globalization – Trade from a Historical Perspective." *Our World in Data*. https://ourworldindata.org/trade-and-globalization

Ortiz-Ospina, Esteban and Diana Beltekian. 2018b. "Trade and Globalization – How Much Do Countries Trade?" *Our World in Data*. https://ourworldindata.org/trade-and-globalization

Parliamentary Commissioner for the Environment. 2000. *New Zealand Under SIEGE: A Review of the Management of Biosecurity Risks to the Environment*. Wellington, New Zealand: Office of the Parliamentary Commissioner for the Environment.

Perkins, John. 2016. *The New Confessions of an Economic Hit Man*. Oakland, CA: Berrett-Koehler Publishers.

Perrings, Charles, Katharina Dehnen-Schmutz, Julia Touza, and Mark Williamson. 2005. "How to Manage Biological Invasions under Globalization." *Trends in Ecology and Evolution* 20(5): 212–215.

Pimental, David. 2002. *Biological Invasions: Economic and Environmental Costs of Alien Plant, Animal, and Microbe Species*. Boca Raton, FL: CRC Press.

Rauh, Virginia, Srikesh Arunajadai, Megan Horton, Frederica Perera, Lor Hoepner, Dana B. Barr, and Robin Whyatt. 2011. "Seven-Year Neurodevelopmental Scores and Prenatal Exposures to Chlorpyrifos, a Common Agricultural Pesticide." *Environmental Health Perspectives* 119(8): 1196–1201.

Robbins, Paul. 2004. "Comparing Invasive Networks: Cultural and Political Biographies of Invasive Species." *Geographical Review* 94(2): 139–156.

Sachs, Jeffrey. 2005. *The End of Poverty? Economic Possibilities for Our Time*. New York: Penguin Press.

Schmitz, Don C. and Daniel Simberloff. 1997. "Biological Invasions: A Growing Threat." *Issues in Science and Technology* 13(4): 33–40.

Schnaiberg, Allan and Kenneth Gould. 1994. *Environment & Society: The Enduring Conflict*. New York: St. Martin's Press.

Spiegelman, Annie. 2010. "The Light Brown Apple Moth (LBAM) Doesn't Deserve the Starring Role." *Huffington Post*. http:// www.huffingtonpost.com/annie-spiegelman/the-light- brown-apple-mot_b_523306.html

Stats NZ. 2021a. "Infoshare: International Travel and Migration: Total Passenger Movements." http://infoshare.stats.govt.nz/ViewTable.aspx?pxID=3b061bdf-7385-4083-9fd0-aca30172d9e3

Stats NZ. 2021b. "Infoshare: International Travel and Migration: Visitor Arrival Totals." http://infoshare.stats.govt.nz/ViewTable.aspx?pxID=548743c6-7abf-449e-9fdd-318eed64fb33

Trading Economics. 2020a. "New Zealand – Trade (% of GDP)." https://trading-economics.com/new-zealand/trade-percent-of-gdp-wb-data.html

Trading Economics. 2020b. "New Zealand Imports: 1960–2020 data." https://tradingeconomics.com/new-zealand/imports

Vilà, Montserrat and Jordi Pujadas. 2001. "Land-Use and Socio-Economic Correlates of Plant Invasions in European and North African Countries." *Biological Conservation* 100: 397–401.

3 Contextualizing the Eradication Response

The New Zealand government responded to PAM's arrival by seeking to eradicate it. This was not an uncommon response as previous New Zealand governments pursued the eradication of other species, including rabbits, the subterranean termite (*Coptotermes acinaciformis*), and the white-spotted tussock moth (*Orgyia thyellina*) (Office of the Ombudsman 2007; Parliamentary Commissioner for the Environment (PCE) 2000). Nor are eradication responses unique to New Zealand as numerous other countries have responded similarly to foreign species. For example, government agencies in the United States have pursued eradication efforts against the Asian gypsy moth, light brown apple moth, Mediterranean fruit fly, zebra mussels, eucalyptus trees, and cogongrass, to name but a few (Cockburn 2015; Gibbons 1992; Gutierrez & Ponti 2011; Schmitz & Simberloff 1997; Spiegelman 2010).

Although eradication is one response to a foreign species, it is not the only one as government agencies can exhibit more tolerant responses, such as merely suppressing, containing, or managing the species in question, with another response being to turn a blind eye to it (Green 2000; Schmitz & Simberloff 1997; Thomas n.d.). For example, while the New Zealand government aggressively pursued PAM's eradication, it merely manages other invasive species, such as stouts, gorse, and rabbits, even though each has significantly impacted indigenous species and local ecosystems (Barker 2008). Additionally, the country has *tolerated* numerous other species that have negatively impacted indigenous species, such as trout, deer, dogs and cats, with the latter two having proven to be particularly devastating to indigenous bird populations (Green 2000). What's more, government officials have actually encouraged importing certain foreign species, such as ornamental plants, exotic pets, or agricultural animals. For example, over the course of British settlement, New Zealand officials actively encouraged importing foreign plant species, which led to an estimated 19,000 to 25,000 foreign plant species being introduced into the country (Brake & Peart 2013; Green 2000). Of those, 2,100 have established in the wild, where they have competed with 2,400 native species (Green 2000). In the 1990s the country clamped down on the importing of known undesirables, though it continued to allow importing non-native plants that were not prohibited (Hulme 2020).

DOI: 10.4324/9780429426414-4

Besides response-variation across species, there are instances where different countries will diverge in their response to the same species. For example, whereas New Zealand responded to PAM with an extensive and expensive eradication campaign, PAM is tolerated in its native Australia, where it is only considered a minor pest on apple trees (Office of the Controller and Auditor-General 2002). Additionally, where California sought to eradicate the light brown apple moth (also from Australia) with an aerial pesticide spraying operation, New Zealand has worked to manage the population through a range of different tools, including biological controls, pheromone traps, and mating disruption (Spiegelman 2010; Suckling & Brockerhoff 2010; Walker et al. 2017). Such variation underscores Foster and Sandberg's (2004) point that responses to species are guided by sociocultural contexts, which need to be unpacked, analyzed, and explained.

Part of that work includes identifying the proponents of eradication campaigns. As Andrew Cockburn (2015) points out, intolerance towards a particular species does not happen in a social vacuum but rather is shaped by powerful forces, such as nativist societies and pesticide manufacturers like Monsanto, who view invasive species as a financial opportunity and have worked to stoke public concerns about and intolerance of new arrivals (Cockburn 2015). In the PAM case, government agencies were quite important, with the Ministry of Agriculture and Forestry (MAF) playing a particularly prominent role. While many other government agencies (including the Auckland District Health Board and the Ministry of Health) provided support to MAF's efforts, MAF was particularly important because they are the agency in charge of biosecurity, they carried out the original eradication response to PAM, and when that response failed, escalated eradication efforts by adding aerial pesticide sprayings in urban areas, which they also oversaw (Office of the Ombudsman 2007). Another important social force was the forestry industry, which felt vulnerable vis-à-vis the moth and communicated its concern to MAF right from the outset (Office of the Controller and Auditor-General 2002). Moreover, when MAF's original response failed to eradicate PAM, industry officials pressured MAF to have its efforts independently reviewed and to start pursuing aerial spraying over suburban neighborhoods. A third important social force in favor of the spraying were the environmental groups, and particularly Forest and Bird, which was New Zealand's oldest environmental group and who strongly supported MAF's decision to pursue aerial pesticide spraying over suburban neighborhoods (MAF 2002b; Royal Forest and Bird Protection Society 2001).

Beyond identifying the social forces pushing for eradication, we also need to embed their efforts in time and space, which includes considering the context's cultural, historical, political economic, and technological configuration. Accordingly, this chapter contextualizes the government's response to PAM by discussing several mediating factors, including the dominant worldview found in industrialized societies, New Zealand's history of struggling with invasive species, the growth-oriented political economy, the country's strong reliance on primary industries, and industrial practices that have stripped ecosystems of biodiversity and have, in turn, rendered them more vulnerable to biological invaders.

The Dominant Worldview in Industrial Societies

It is essential to start by considering the dominant worldview, as the worldview a society adheres to will shape how its people view the environment and the different species within it, how they relate to those species, and, in turn, how they interact with them. Charles Harper (2001) defines worldviews as "the totality of cultural beliefs and beliefs systems about the world and reality that people share" (p. 38). Moreover, he identifies the worldviews for three main types of society (hunter and gatherer, agricultural, and industrialized). Each worldview has a central organizing theme that conceptualizes nature and the relation of people to it, which, in turn, influences human behaviors.

The central theme in the hunter and gatherer worldview is that humans are embedded in the natural world. They conceptualize the environment as a "*living natural world* (wilderness/jungle/forest/ grassland) of things and beings governed by spiritual forces" (Harper 2004, p. 40). Moreover, they see themselves as people in nature, who "survived by being keenly aware of their dependence on nature and each other" (*Ibid.*). That does not mean they are immune to environmentally destructive behavior, such as driving numerous large game off of cliffs in order to harvest a few. However, it is to say their central orientation was to be attentive to their ecosystems and to endeavor to live within them, as their survival depended on it.

This contrasts profoundly with agricultural societies, where the onus is not on living *with* nature but rather dominating and controlling it (Harper 2004). In these societies, nature is viewed as a garden, which humans are responsible for clearing, plowing, watering, and tending. As part of their tending work, humans eliminate species that do not conform to their vision for their garden, such as wild animals that compete with livestock, wild plants that interfere with planted crops, and insects that feast upon crops (Harper 2004). An important consequence of this worldview was reduced resilience against biological invaders, as monocultures reduce an ecosystem's biodiversity, which creates more opportunities for invasive species to establish a foothold (Schmitz & Simberloff 1997). Moreover, as agricultural societies grew, an increasing proportion of wilderness was taken over by monocultural agriculture and urban environments, both of which spread reduced resilience to larger swathes of land.

In industrial societies the impulse to control and dominate nature is pushed even further. Beyond viewing nature as a garden to be tended, industrial people view it as a large resource base to be manipulated and managed to satisfy human needs and desires (Harper 2012). As well, over the twentieth century, nature has increasingly been conceptualized as a factory, whose outputs need to be maximized through the manipulation of inputs (i.e. labor, control over

the species that are allowed to live, chemical pesticides, and chemical fertilizers). In this worldview there is little value given to nature for its own sake,
and it is assumed that technological inventiveness will be able to overcome all
physical limits to growth.

The New Zealand Case

Considering a country's dominant worldview provides a useful handle on the
New Zealand case. First, the agricultural worldview corresponds well with
the way settlers viewed and transformed the New Zealand landscape (Bell
et al. 2017). Maori and British settlers both pursued significant deforestation,
though British settlers pursued it for the specific purpose of increasing agricultural land (Anderson 2013; Dann 2002; Wynn 2013). In their pursuit of
agricultural land, British settlers aggressively drained ecologically prodigious
wetlands, reducing by 85%–90% the wetlands that existed prior to British
colonization (Gerbeaux 2003; Park 2013; Taylor et al. 1997). This loss of
wetlands represents one of the greatest rates of wetland loss in the developed
world (Mitsch & Gosselink 2007). Besides transforming the landscape into
gardenscape, European settlers also aggressively modified the ecological content of those spaces, importing a wide range of birds, mammals, and plants
into the New Zealand context. For instance, since the beginning of British
colonization, in 1840, settlers have introduced an estimated 19,000 to 25,000
plant species (Brake & Peart 2013; Green 2000).

With the industrialization of society, New Zealand increasingly viewed
farmland as factories, whose outputs needed to be maximized by monocropping and relying on artificial fertilizers (Brooking & Star 2011). This
tendency grew significantly in the post-World War I (WWI) era, as the
government increasingly shaped agricultural practice to expand productivity (*Ibid.*). Additionally, after WWII New Zealanders embraced the use of
chemical insecticides and herbicides to eliminate unwanted insects and plants
(Dann 2002; OECD 1996; Rolando et al. 2016; Watts 1994).

A similar dynamic occurred with forestry. After WWI a growing timber
shortage created a drive for reforesting, which led to trees being planted on
large tracts of land (Roche 2013). For example, during 1921–1922 the State
Forest Service (SFS) planted trees on 3,400 acres (*Ibid.*). Moreover, by 1934 SFS
had planted trees on over 400,000 acres, with private companies planting on
another 270,000 acres (*Ibid.*). Importantly, these plantations were mainly monocultures, which became increasingly dominated by the fast-growing Monte-
rey Pine (*Pinus radiata*), a foreign import that was first brought to New Zealand
in the 1860s, and which, over the last few decades, has made up 90% of New
Zealand plantations (Maclaren 1996; Ministry of Forestry 1993; Roche 2013).

Importantly, the high monoculture rate introduced two dynamics that
increased the industry's vulnerability to foreign species. The first is that increased monocultures reduced natural forest biodiversity (i.e. a reduction in
tree, bird, insect, and fungi species), which normally serves as a natural buffer

against foreign species. Additionally, the forestry industry increasingly relied on chemical herbicides to eliminate unwanted plants, which further reduced forest biodiversity and the natural protection it provides against invasive species (Coster et al. 1986; Rolando et al. 2016).

As this relates to the PAM incursion, in the industrial worldview any species that interferes with productivity is seen as a threat that needs to be exterminated. Given that feeding trials suggested PAM *could* feed on pine trees when their preferred food source is unavailable and that this could marginally reduce industry profitability, we should have expected that biosecurity agents would target PAM for eradication. This is particularly true when we consider that monocropping and pesticide use had produced ecosystems that were more vulnerable to a new species.

Aotearoa New Zealand's Long History of Struggling with Foreign Species

Another important contextual element was New Zealand's long history of struggling with foreign species, which included having to deal with the ecological and economic consequences of those struggles. This began with Polynesian settlement, which introduced rats and dogs, both of which devastated indigenous species (Anderson 2013). Biological invaders became even more of a problem with British imperialism, which encouraged settlers to import new species for recreational, sentimental, economic, and productive reasons, and which had destructive effects on indigenous flora, fauna, and ecosystems. This was no different in Aotearoa New Zealand, where settlers introduced cats, dogs, mustelids (i.e. ferrets, weasels, and stouts), possums, and two new species of rats (joining the one introduced by Polynesian settlers) (Dann 2002). Each of these introductions contributed to the decimation of kiwis and other native birds, who evolved in the absence of large mammals and do not have effective defenses against predators. Indigenous land-based birds have been particularly affected, with 40 now extinct and 37 of the remaining 51 classified as "threatened" (Wilson 2015).

Besides undermining local fauna, numerous imported species (such as rabbits, deers, mustelids, and possums) have also engendered significant economic costs, with rabbits being a particularly problematic import. Rabbits were first introduced in the 1830s for food and sport but, without natural predators to keep them in check, grew to plague proportions on several occasions, including the 1870s, 1920s, 1940s, and 1980s (Isern 2002; Peden 2008). The rabbit population's explosive growth destroyed vegetation cover and triggered soil erosion, which reduced the ecosystem's capacity to support sheep raising (Peden & Holland 2013). The rabbits' environmental impacts have been particular costly to farmers, and in 1999 production losses were estimated to be around $50 million per year (Peden 2008). Additional costs have been incurred through the various activities that governments have pursued to contain the problem, which included passing legislation (several Rabbit Nuisance Acts), establishing the national Rabbit Destruction Board in 1947, developing Task Forces to study the problem, hiring professional hunters, and administering the 1080 poison (Isern 2002).

An important knock-on effect was that, in an effort to curtail the rabbit population, in the 1870s New Zealand began importing mustelids (ferrets, stoats, and weasels). Mustelids were considered a natural predator of rabbits and, thus, a tool to biologically control the latter. Unfortunately, however, that solution was not particularly effective, as underscored by a surge in rabbit populations in the 1920s, 1940s, and 1980s (Isern 2002; Peden 2008). Moreover, importing mustelids created an additional biosecurity problem, as the lack of natural predators allowed mustelid populations to also grow prodigiously, which, as some predicted in the 1870s, also became a scourge to indigenous lizards and ground-dwelling birds (Peden & Holland 2013).

Another import that has plagued Aotearoa is the Australian brushtail possum, which the British originally brought to the country in 1858 to establish a fur trade (Isern 2002). That trade began to grow in the 1870s, which led to a growth in the possum population. At its peak, in the 1980s, their population was estimated to be between 60 and 70 million (Isern 2002; Warburton et al. 2009). Possums have ravaged native flora, collectively consuming about 10 tons of vegetation per night, most of which is new growth (Nugent n.d.). As well, they have been a menace to local fauna, consuming eggs, birds, and bats (Brockie 2015). Beyond its impact on native flora and fauna, the possum problem is estimated to cost farmers $35 million a year, which has compelled the New Zealand government to annually spend $80 to $110 million on possum control (Predator Free NZ n.d.; Warburton et al. 2009).

Beyond importing fauna, British settlers also imported problematic plants, such as gorse (also known as ulex, furze, or whin), which is a thorny evergreen shrub that English settlers brought to Aotearoa in the early 1800s to use as hedges. In the temperate New Zealand climate the shrub spread like wildfire, to the point local governments began viewing it as a pest in the early 1860s (Dawson 2010). It now covers over 1,700,000 acres of New Zealand land, with much of it existing in hill country, which reduces the grazing area for livestock (Blaschke et al. 1981). It also competes significantly with young indigenous trees for space, and its thorns makes forestry operations (such as thinning and pruning) more difficult and costly (Massey University 2019). Aside from having thorns that make it difficult to remove, it is seemingly impervious to burning or bulldozing, both of which create ideal conditions for its seeds to disperse and germinate (Blaschke et al. 1981). Many consider it to be New Zealand's worst weed, and in the early 1980s the government spent over $22 million a year to control it (Timmins 1988). As well, between 1981 and 1985 a government subsidy program encouraged farmers to use 2,4,5-T (a particularly toxic herbicide laced with carcinogenic dioxins) on the weed, which led farmers to spray millions of liters over the New Zealand countryside (Takoko & Gibbs 2004). Beyond the example of gorse, an estimated 19,000 to 25,000 plants have been introduced into New Zealand, with over 240 classified as invasive weeds (Brake & Peart 2013; Green 2000).

Although the country has not struggled as much with insect invaders, there have been a few attention-grabbing cases, such as the Asian tiger mosquito (*Aedes albopictus*), which is from Southeast Asia. In the mid-1980s this mosquito spread globally via containers filled with used tires and was a concern because it carried 18 viral diseases, including the potentially fatal dengue fever (Green 2000). By 2000 the mosquito had established itself in numerous new locales, including the Southeastern USA, Southern Europe, Nigeria, South Africa, Brazil, Australia, and briefly New Zealand (Green 2000; Laird et al. 1994). Another attention-grabbing case was the southern saltmarsh mosquito (*Aedes camptorhynchus*), which is from Australia. This mosquito is a vector for the debilitating Ross River virus and had established itself in Napier city, a small city in Eastern New Zealand (PCE 2000). Besides these two species, New Zealand border control agents have intercepted 16 other species of exotic mosquitos (Derraik 2004).

Besides mosquitos, another insect invader has been the Argentine ant (*Linepithema humile*), which was discovered at the site of 1990 Auckland Commonwealth Games and within a decade had managed to spread throughout the country (PCE 2000). This species was a biosecurity concern because of its documented ability to outcompete and displace native insects in other countries, which has triggered important ecosystem effects. For example, in South Africa it displaced native ant species, which led to the collapse of plant species that relied on the indigenous ants to disperse and bury its seeds (PCE 2000).

Given New Zealand's long history of struggling with invasive species and the ecosystem and economic impacts they have engendered, the country has become increasingly sensitized to the potential impact of invasive species. Moreover, this increased sensitivity resulted in passage of the 1993 Biosecurity Act, which enhanced the country's biosecurity apparatus, by establishing an integrated system of biosecurity risk management that included clearly defined biosecurity responsibilities (central government was responsible for border control, whereas regional governments were responsible for pest management functions) and infrastructure to monitor and manage (i.e. control or eradicate) unwanted organisms (PCE 2000). This integrated approach was the first of its kind in the world (Learnz n.d.; MAF 2009) and gave government agencies the mandate and power to aggressively seek out and eradicate unwanted foreign species. In turn, this proved important to the painted apple moth story, for when the moth was discovered in Auckland in 1999, the government had the legal and logistical structures in place to aggressively seek out and eliminate it.

Although dominant worldviews and the country's history of struggling with invasive species offer some explanatory power for the government's desire to eradicate PAM, by themselves these factors are incomplete, as they do not explain why the government would go after PAM while tolerating thousands of other invasive species, some of which are a recognized threat to indigenous species (Green 2000; PCE 2000). For example, there are over 2,100 unwanted foreign plant species in the country that are competing with

2,400 native species, yet they have not been the subject of eradication operations. The point is also underscored by the case of Argentine ants, which have demonstrated the ability to outcompete native insects in many countries, but which the New Zealand government did not seek to eradicate (Green 2000; PCE 2000).

For a deeper understanding of why governments pursue the eradication of some invasive species but not others, it is important to situate those decisions in their political and economic context, as the political economic configuration will shape whose interests are prioritized, and when government bureaucracies activate their biosecurity apparatus.

Relating Biosecurity Actions to Political Economy

A political economy of the environment frameworkcalls for closely linking government actions to the configuration of power. For instance, if corporations are the dominant social power of the day, then one expects to see government policies, discourses, and actions to be heavily tilted towards protecting industry interests, such as setting weak environmental regulations, underfunding the enforcement of those regulations, and even undertaking ecological vandalism on behalf of industry (Gould et al. 2008; Schnaiberg & Gould 1994). Thus, to understand government actions, it is crucial to analyze and understand the political economy within which those actions are taking place. Moreover, political economies do not emerge in a social vacuum, but rather evolve over time, in conjunction with world orders. So, it is also important to embed political economies within their historical and political contexts.

This section traces the evolution of the New Zealand political economy, emphasizing the nation's initial status as a colony in the British Empire, its economy's orientation towards primary industries, and how that orientation made it particularly susceptible to potential biological invaders.

A Strong Reliance on Primary Industries

As a former colony in the British Empire, New Zealand's governance was directed by the imperial center, which was more than 11,000 miles away. A problem with colonial governance structures is that decisions tend to benefit the imperial center, instead of prioritizing the well-being of local populations (Wallerstein 2011). One way to benefit the center is to orient colonial economies towards extractive industries, which will send raw resources to the center for processing, instead of establishing local economies that will prioritize the enrichment and well-being of local populations (Davis 2010; Diaz 2010; Perkins 2016). As was the case in the other British colonies, the British Crown oriented New Zealand's economy towards primary industries, which included mining, logging, and agriculture (Dann 2002; Pawson & Brooking 2013).

As New Zealand's economy grew it continued to rely significantly on primary industries. Although we are now said to be living in a "post-colonial" world order, a legacy of the colonial world order is that the political economies of many former colonies are still oriented towards producing primary materials for export (Diaz 2010; Perkins 2016). While particularly true for developing countries, it is also true for settler countries like New Zealand. For instance, even though New Zealand gained its independence in 1947, it has retained a strong orientation towards primary production for export markets, with agriculture and forestry being particularly important to the country's economy (Wilson 2006). In 1972 agriculture's share of GDP reached a high of 13.86%. By 1990 the percentage had dropped to 7.59%, though it stayed relatively stable throughout the 1990s, remaining between 6.95% and 9.09% of GDP. In 1999, the year of PAM's arrival, agriculture represented 7.5% of the New Zealand's GDP, with forestry contributing an additional 3% (Dann 2002; The Global Economy 2020a). Primary industries were also significant contributors to exports. For instance, in 1998/99 the agriculture, horticulture, fishery, and forestry sectors accounted for 70% of export receipts, which represented approximately $15.3 billion (Rauniyar et al. 1999).

To better understand primary industry's importance to the New Zealand economy, it is instructive to compare agriculture's share of GDP among industrialized democracies. Figure 3.1 features the five industrialized democracies where agriculture represented the greatest share of GDP in the decade leading up to PAM's 1999 arrival and shows that New Zealand considerably outpaced its nearest competitors. The one exception was 1995, when Greece's figure nearly matched New Zealand's (7.38% versus 7.57%) (*Ibid.*) after which New Zealand significantly widened the gap. Moreover, in 2000 its figure was nearly three times that of Australia, which was the other settler country on the list (The Global Economy 2020b).

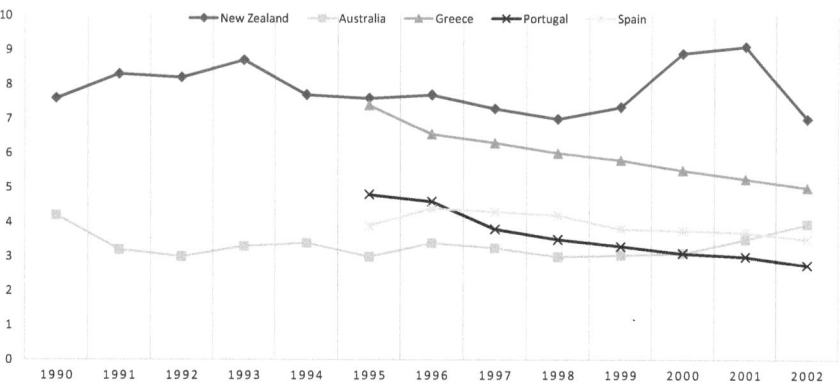

Figure 3.1 Contribution of Agricultural Sector to GDP, as % of GDP (The Global Economy 2020b).

Besides contributing to exports and GDP, primary industries also con-tribute significantly to employment. For instance, in the decade preceding the PAM campaign the New Zealand agricultural sector provided between 10.9% (in 1991) and 8.7% (in 2000) of New Zealand's total employment and in 1996 employed 154,665 people (Fairweather et al. 2000; The Global Economy 2020c). Regarding the forestry industry, in 1996 it employed over 25,000 people directly and many more indirectly (Fairweather et al. 2000).

Given New Zealand's long history of struggling with invasive species and its economy's significant reliance on primary industries, government officials have been wary of biological invaders that could impact primary industries. This is underscored by the fact most biosecurity funding has been allocated to controlling invasive species that could threaten primary industries, such as agriculture and forestry. Specifically, 95% of the 1999 biosecurity funding was allocated to the Ministry of Agriculture and Forestry (which is responsible for eliminating species that threaten agriculture and/or forestry), with only 5% going to the Department of Conservation (which is responsible for controlling invasive species that might threaten indigenous species) (PCE 2000).

As this pertains to PAM, in its native Australia the moth is viewed as a minor garden pest (Suckling & Brockerhoff 2010), and that modest impact could have tempered concerns about its potential impact in New Zealand. Moreover, while the moth was known to cause some defoliation of pine trees in Australia, the defoliation only occurred for younger trees, only slowed growth (instead of killing them), and was relatively minor in extent (less than 1% of trees in infested sections) (MAF 2000).

However, in New Zealand, just its status as a foreign species would have been enough to concern government officials. Additionally, those concerns would have been exacerbated by the fact that *Radiata pine* is one of the trees PAM *could* feed on. Even though pine trees were not the moth's preferred food source, the potential it *might* feed on those trees would have prompted government officials to view the moth as a potentially significant economic risk for the country, as pine trees represented 90% of trees in New Zealand plantation forests (MAF 2000). These economic concerns are reflected in MAF's (2000) initial economic assessment of the incursion, which predicted that failing to eradicate PAM would cost the country between $16 and $116 million, with the likely final total being around $48 million. Interestingly, that estimated impact was not distributed across many industries but rather focused on the forestry industry, which MAF analysts predicted would shoul-der 78% to 87% of the economic impact (MAF 2002a).

It is also interesting to note that the estimated economic impact did not reflect predicted increases in operating costs, but rather the predicted lost profits that would emerge *if* tree infestation rates were 2–25 times higher than they typically are for pine tree plantations in Australia. Such question-able assumptions are something I explore more deeply in Chapter 7, where I discuss how the government sought to build support for the aerial spraying by framing PAM as a triple biosecurity threat.

The Increased Vulnerability Caused by Industrialization

Besides New Zealand's reliance on primary industries, the country's vulnerability to biological invaders was exacerbated by the industrialization of those economic sectors, which weakened ecosystem biodiversity and, in turn, resilience against biological invaders.

Industrialization in New Zealand began with the Taylorization of sheep shearing, where the activity was broken down into its sub-activities and each worker was allocated responsibility for one of those sub-activities (Dann 2002). Then agricultural was similarly Taylorized (*Ibid.*). The next stage of industrialization was the mechanization of production, which began with the use of primitive coal and water-powered machines, which were later replaced by machinery relying on fossil fuels, such as tractors and harvesters (*Ibid.*). Another important aspect of industrialization was the increased reliance on synthetic chemicals, starting with synthetic fertilizers in post-WWI era and then synthetic pesticides in the post-WWII era (Dann 2002; OECD 1996; Rolando et al. 2016; Watts 1994).

Still another aspect of the industrialization process was industry's growing tendency to conceptualize nature as a factory, whose crops were to be manipulated to maximize short-term outputs and profits. John Bellamy Foster (1999) refers to this practice as the "scientific management of nature" and an example he provides is the management of plantation forests, which tends to favor fast-growing species (such as pine trees) and the fastest growing genetic variant of that species. New Zealand provides a prime example of this concept as the forestry industry has demonstrated a proclivity for using the fast-growing *Pinus radiata*, which typically makes up about 90% of the trees in the country's plantation forests (Maclaren 1996; Ministry of Forestry 1993).

A problem with the scientific management of nature is that it drastically reduces the species and genetic diversity of trees, thereby making plantations more vulnerable to foreign species and/or other threatening biota. In turn, such vulnerabilities lead plantation owners to rely heavily on pesticides. Although New Zealand forestry has not been a big user of insecticides (Maclaren 1996), it has been a significant user of chemical herbicides to control for weeds (Coster et al. 1986). In the mid-1970s plantation owners used chemicals for 66% of post-planting weed control (*Ibid.*). Moreover, they relied extensively on 2,4,5-T (one of two principle chemicals in Agent Orange), which was the industry's main weed control from the mid-1950s until it was banned in 1989 (*Ibid.*). In the first part of the 1980s New Zealand used vast amounts of this pesticide, applying 57,000 lbs of it in 1985 alone (Coster et al. 1986, p. 48). After 2,4,5-T was banned plantation operators turned to 2,4-D (i.e. the other half of the Agent Orange formulation), as well as a host of other herbicides, including picloram, glyphosate, hexazinone, triclopyr, paraquat, diquat, and atrazine (Rosoman et al. 1994).

A problem with relying on chemical pesticides is that they act indiscriminately, which has numerous consequences for ecosystem health (Pimentel

2005; Pimentel et al. 1992). First, beyond killing the target species, chemical pesticides harm beneficial species. For example, New Zealand forestry's use of 2,4-D, glyphosate, hexazinone and triclopyr has been shown to adversely affect ectomycorrhizal fungi, which is key to the health of trees as it increases nutrient uptake, improves resistance to stress in trees, and is particularly beneficial to *Pinus radiata*, the predominant species in New Zealand forestry (Rosoman et al. 1994). Chemical pesticides also kill beneficial insects (such as spiders, ladybugs, and parasitic wasps), which would otherwise help minimize the impact of foreign species. In turn, the loss of beneficial insects sets in motion the "pesticide treadmill" (Magdoff et al. 2000), whereby increasing the ecosystem's vulnerability to biological invaders further increases the manager's reliance on pesticides to ward off that threat, which only further weakens the ecosystem's natural resilience and further escalates the manager's reliance on pesticides.

Another problem with pesticides is that they harm birds, decreasing their numbers and reducing their ability to protect an ecosystem from insect infestations. Insect-eating birds have been found to be particularly absent in plantation forests (Rosoman et al. 1994). Decreasing bird populations also speed up the pesticide treadmill, as their diminished numbers further escalate the plantation forest's susceptibility to infestations, thereby increasing even more the plantation manager's reliance on pesticides.

Industrializing the forestry industry has increased its vulnerability to invasive species, which undoubtedly increased industry and government sensitivity to the arrival of foreign species. Beyond having a general vulnerability to foreign species, the New Zealand forestry industry was particularly concerned about the painted apple moth, due to the belief pine trees were a species of tree the moth *could* feed on (MAF 2000, 2002). This concern was exacerbated by the fact MAF mishandled the initial biosecurity response for PAM, thereby allowing the moth to spread considerably in its first two and a half years in New Zealand. The level of industry concern is indicated by the fact industry leaders openly questioned MAF's management of the PAM incursion and pressed for an independent review of it, which they obtained in May 2001, when MAF commissioned such a review (Dyck 2001; OAG 2002).

Summary

In trying to account for a government's response to an invasive species there are many factors to consider, starting with the society's dominant worldview, which will shape how humans view nature and their place in it, both of which will, in turn, influence how humans interact with different species. In the dominant worldview of industrialized societies nature is viewed as a resource to be managed and dominated to satisfy human needs and desires. Part of that "management" includes eliminating species that could impact industry profits.

A second factor to consider is the nation's history of battling against invasive species. The greater the struggles, the less likely governments are to tolerate the arrival of new species, and the more likely they are to have a bureaucracy that will enable them swiftly respond to unwanted species, as has been the case in New Zealand.

Third, we also need to consider the political economy in which the nation is embedded. In a capitalist political economy, governments will be oriented towards protecting industrial interests, which includes eliminating foreign species that might interfere with primary industries. This will be particularly true with an economy that relies significantly on primary industries, as has been the case in New Zealand. Further enhancing their sensitivity to biological invaders is the degree to which industries are practicing the "scientific management of nature," which is associated with biodiversity loss and increased vulnerability to biological invaders.

While this chapter's analysis helps elucidate why the New Zealand government strove to eradicate the painted apple moth, it does not explain why they sought to do so with an extensive aerial pesticide spraying operation. Illuminating that issue will be the next chapter's focus.

References

Anderson, Atholl. 2013. "A Fragile Plenty: Pre-European Maori and the New Zealand Environment." In *Making a New Land: Environmental Histories of New Zealand*, edited by Eric Pawson and Tom Brooking, 35–51. Dunedin, NZ: Otago University Press.

Barker, Kezia. 2008. "Flexible Boundaries in Biosecurity: Accommodating Gorse in *Aotearoa* New Zealand." *Environment and Planning A* 40: 1598–1614.

Bell, Avril, Vivienne Elizabeth, Tracey McIntosh, and Matt Wynyard. 2017. *A Land of Milk and Honey?: Making Sense of Aotearoa New Zealand*. Auckland, NZ: Auckland University Press.

Blaschke, Paul M., Grant G. Hunter, G. O. Eyles, and P. R. van Berkel. 1981. "Analysis of New Zealand's vegetation cover using land resource inventory data". *New Zealand Journal of Ecology* 4: 1–19

Brake, Lucy and Raewyn Peart. 2013. *Treasuring our Biodiversity: An EDS Guide to the Protection of New Zealand's Indigenous Habitats and Species*. Auckland, NZ: Environmental Defence Society.

Brockie, Bob. 2015. "Introduced animal pests – Possums." *Te Ara – The Encyclopedia of New Zealand*. https://teara.govt.nz/en/introduced-animal-pests/page-2

Brooking, Tom and Paul Star. 2011. "Remaking the Grasslands: The 1920s and 1930s." In *Seeds of Empire: The Environmental Transformation of New Zealand*, edited by Tom Brooking and Eric Pawson, 178–199. London, UK: Palgrave Macmillan.

Cockburn, Andrew. 2015, September. "Weed Whackers: Monsanto, glyphosate, and the war on invasive species." *Harper's Magazine*, pp. 57–63.

Coster, A. P, A. S. Edmonds, J. M. Fitzsimons, C. G. Goodrich, J. K. Howard, and J. R. Tustin. 1986. *The Use of 2,4,5-T in New Zealand: A Report to the Environmental Council*. A Report by a Working Party to the Environmental Council, Wellington: Commission for the Environment for the Environmental Council.

Dann, Christine. 2002. "Losing Ground? Environmental Problems and Prospects at the Beginning of the Twenty-First Century." In *Environmental Histories of New Zealand*, edited by Eric Pawson and Tom Brooking, 275–287. Auckland, NZ: Oxford University Press.

Davis, Mike. 2010. *Late Victorian Holocausts: El Niño Famines and the Making of the Third World*. Brooklyn, NY: Verso Press.

Dawson, Bee. 2010. *A History of Gardening in New Zealand*. Auckland, NZ: Godwit.

Derraik, Jose. 2004. "Exotic Mosquitoes in New Zealand: A Review of Species Intercepted, Their Pathways and Ports of Entry." *Australian New Zealand Journal of Public Health* 28: 433–44.

Diaz, Philippe. 2010. *The End of Poverty? Think Again*. Canoga Park, CA: Cinema Libre Studio.

Dyck, Bill. 2001. "Painted Apple Moth Response under Scrutiny." *Biosecurity New Zealand* 28: 13.

Fairweather, John R., Peter J. Mayell, and Simon R. Swaffield. 2000. *A Comparison of the Employment Generated by Forestry and Agriculture in New Zealand*. Research Report No. 246. Canterbury, NZ: Agribusiness and Economics Research, Lincoln University.

Foster, Jennifer and L. Anders Sandberg. 2004. "Friends or Foe? Invasive Species and Public Green Space in Toronto." *Geographical Review* 94(2): 178–198.

Foster, John Bellamy. 1999. *The Vulnerable Planet: A Short Economic History of the Environment*. New York: Monthly Review Press.

Gerbeaux, Philippe. 2003. "The Ramsar Convention: A Review of Wetlands Management in New Zealand." *Pacific Ecologist* 4: 37–41.

Gibbons, Ann. 1992. "Asian Gypsy Moth Jumps Ship to United States." *Science* 255(5044): 526.

Gould, Kenneth A., David N. Pellow, and Allan Schnaiberg. 2008. *The Treadmill of Production*. Boulder, CO: Paradigm Publishers.

Green, Wren. 2000. *Biosecurity Threats to Indigenous Biodiversity in New Zealand – An Analysis of Key Issues and Future Options*. A Background report prepared for the Parliamentary Commissioner for the Environment, Wellington.

Gutierrez, Andrew Paul and Luigi Ponti. 2011. "Assessing the Invasive Potential of the Mediterranean Fruit Fly in California and Italy." *Biological Invasions* 13: 2661–2676.

Harper, Charles. 2001. *Environment and Society: Human Perspectives on Environmental Issues, 2nd edition*. Upper Saddle River, NJ: Prentice Hall.

Harper, Charles. 2004. *Environment and Society: Human Perspectives on Environmental Issues, 3rd edition*. Upper Saddle River, NJ: Prentice Hall.

Harper, Charles. 2012. *Environment and Society: Human Perspectives on Environmental Issues, 5th edition*. Boston, MA: Prentice Hall.

Hulme, Philip. 2020. "Plant Invasions in New Zealand: Global Lessons in Prevention, Eradication and Control." *Biological Invasions* 22: 1539–1562.

Isern, Thomas. 2002. "Companions, Stowaways, Imperialists, Invaders: Pests and Weeds in New Zealand." In *Environmental Histories of New Zealand*, edited by Eric Pawson and Tom Brooking, 233–245. Melbourne, AU: Oxford University Press.

Jay, Mairi and Munir Morad. 2006. "The Socioeconomic Dimensions of Biosecurity: The New Zealand Experience." *International Journal of Environmental Studies* 63(3): 293–302.

Laird Marshall, Lester Calder, Richard C. Thornton, Rachel Syme, Peter W. Holder, and Motoyoshi Mogi. 1994. "Japanese *Aedes albopictus* among Four Mosquito

Species Reaching New Zealand in Used Tires." *Journal of the American Mosquito Control Association* 10(1): 14–23.

Learnz. n.d. "Biosecurity." *New Zealand Primary Industries*. http://www.learnz.org.nz/primaryindustries172/bg-standard-f/biosecurity

Maclaren, J. Piers. 1996. *Environmental Effects of Planted Forests in New Zealand: The Implications of Continued Afforestation of Pasture.* Rororua, NZ: New Zealand Forest Research Institute.

Magdoff, Fred, John Foster, and Fred Buttel. 2000. *Hungry for Profit: The Agribusiness Threat to Farmers, Food, and the Environment.* New York: Monthly Review Press.

Massey University. 2019. "Gorse, Botanical name: Ulex europaeus." *Weeds database,* Massey University. https://www.massey.ac.nz/massey/learning/colleges/college-of-sciences/clinics-and-services/weeds-database/gorse.cfm

Ministry of Agriculture and Forestry (MAF). 2000. *Potential Economic Impact on New Zealand of the Painted Apple Moth,* July 2000.

Ministry of Agriculture and Forestry (MAF). 2002a. *Painted Apple Moth: Reassessment of Potential Economic Impacts,* May 2002.

Ministry of Agriculture and Forestry (MAF). 2002b. "MAF Plans to Abandon Painted Apple Moth Project." *Scoop Independent News,* June 24. https://www.scoop.co.nz/stories/PO0206/S00156/maf-plans-to-abandon-painted-apple-moth-project.htm?from-mobile=bottom-link-01

Ministry of Agriculture and Forestry (MAF). 2009. *Review of Key Parts of the Biosecurity Act 1993.* Wellington, NZ: Ministry of Agriculture and Forestry. www.biosecurity.govt.nz/biosec/pol/biosecurity-act-review

Ministry of Forestry. 1993. *New Zealand Forestry Statistics. April 1993.* Ministry of Forestry: Wellington, NZ.

Mitsch, William J. and James G. Gosselink. 2007. *Wetlands, 4th edition.* New York, John Wiley.

Nugent, Graham. n.d. "Possums… Their Effects on Native Vegetation." *Landcare Research.* https://www.landcareresearch.co.nz/__data/assets/pdf_file/0006/42000/possum_native_vege.pdf

OECD. 1996. *Environmental Performance Reviews: New Zealand.* Paris: Organization for Economic Co-operation and Development.

Office of the Controller and Auditor-General (OAG). 2002. *Report of the Controller and Auditor-General: Management of Biosecurity Risks: Case Studies.* Wellington, NZ: The Audit Office. https://oag.parliament.nz/2002/biosecurity-case-studies/docs/part3.pdf

Office of the Ombudsman. 2007. *Report of the Opinion of Ombudsman Mel Smith on Complaints Arising from Aerial Spraying of the Biological Insecticide Foray 48B on the Population of Parts of Auckland and Hamilton to Destroy Incursions of Painted Apple Moths, and Asian Gypsy Moths, Respectively.* Wellington, NZ: Office of the Ombudsman.

Park, Geoff. 2013. "'Swamps Which Might Doubtless Easily Be Drained': Swamp Drainage and Its Impact on the Indigenous." In *Making a New Land: Environmental Histories of New Zealand,* edited by Eric Pawson and Tom Brooking, 174–189. Dunedin, NZ: Otago University Press.

Parliamentary Commissioner for the Environment. 2000. *New Zealand under Siege: A Review of the Management of Biosecurity Risks to the Environment.* Wellington, NZ: Office of the Parliamentary Commissioner for the Environment.

Pawson, Eric and Tom Brooking. 2013. "Introduction." In *Making a New Land: Environmental Histories of New Zealand,* edited by Eric Pawson and Tom Brooking, 17–31. Dunedin, NZ: Otago University Press.

Peden, Robert. 2008. "Rabbits." *Te Ara – The Encyclopedia of New Zealand.* https://teara.govt.nz/en/rabbits/print

Peden, Robert and Peter Holland. 2013. "Settlers Transforming the Open Country." In *Making a New Land: Environmental Histories of New Zealand*, edited by Eric Pawson and Tom Brooking, 89–105. Dunedin, NZ: Otago University Press.

Perkins, John. 2016. *The New Confessions of an Economic Hit Man.* Oakland, CA: Berrett-Koehler Publishers.

Pimentel, David. 2005. "Environmental and Economic Costs of the Application of Pesticides Primarily in the United States." *Environment, Development and Sustainability* 7: 229–252.

Pimentel, David, H. Acquay, M. Biltonen, P. Rice, M. Silva, J. Nelson, V. Lipner, S. Giordano, A. Horowitz, and M. D'Amore. 1992. "Environmental and Economic Costs of Pesticide Use." *American Institute of Biological Sciences* 42(10):750–760.

Predator Free NZ. n.d. "Possum Facts." https://predatorfreenz.org/resources/introduced-predator-facts/possum-facts/

Rauniyar, G. P., A. S. W. Winton, C. F. Whyte, and J. Cheyne. 1999. *Assessment of the Effects of Border Strategies at Modifying Aircraft Passenger Behaviour.* MAF Policy Project No. MSH 301. Massey University, Palmerston North.

Roche, Michael. 2013. "An Interventionist State: 'Wise Use' Forestry and Soil Conservation." In *Making a New Land: Environmental Histories of New Zealand*, edited by Eric Pawson and Tom Brooking, 209–225. Dunedin, NZ: Otago University Press.

Rolando, Carol, Brenda Baillie, Toni Withers, Lindsay Bulman, and Loretta Garrett. 2016. "Pesticide Use in Planted Forests in New Zealand." *New Zealand Journal of Forestry* 61(2): 3–10.

Rosoman, Grant, Greenpeace New Zealand, Canterbury Branch of the Mariua Society. 1994. *The Plantation Effect: An Ecoforestry Review of the Environmental Effects of Exotic Monoculture Tree Plantations in Aotearoa/New Zealand.* Auckland, NZ: Greenpeace New Zealand. http://www.redd-monitor.org/wp-content/uploads/2015/08/The_Plantation_Effect.pdf

Royal Forest and Bird Protection Society. 2001. "Spread of moth is a biosecurity emergency." *Scoop Independent News*, December 3rd. https://www.scoop.co.nz/stories/AK0112/S00006.htm

Schmitz, Don and Daniel Simberloff. 1997. "Biological Invasions: A Growing Threat." *Biological Invasions* 13(4): 33–40.

Schnaiberg, Allan and Kenneth Gould. 1994. *Environment and Society: The Enduring Conflict.* Caldwell, NJ: Blackburn Press.

Spiegelman, Annie. 2010. "The Light Brown Apple Moth (LBAM) Doesn't Deserve the Starring Role." *Huffington Post.* http:// www.huffingtonpost.com/annie-spiegelman/the-light- brown-apple-mot_b_523306.html

Suckling, David Maxwell and E. G. Brockerhoff. 2010. "Invasion Biology, Ecology, and Management of the Light Brown Apple Moth (Tortricidae)." *Annual Review of Entomology* 55: 285–306.

Takoko, Mere and Andrew Gibbs. 2004. *Nga Matitapu o te Hakino: People Poisoned Daily.* Auckland: Greenpeace NZ.

Taylor, Rowan, Ian Smith, Peter Cochrane, Brigit Stephenson, and Nicci Gibbs. 1997. *The State of New Zealand's Environment 1997.* Wellington, NZ: Ministry for the Environment. ISBN 0-478-09000-5.

The Global Economy. 2020a. "New Zealand: GDP share of agriculture." *The Global Economy.* https://www.theglobaleconomy.com/New-Zealand/Share_of_agriculture/

The Global Economy. 2020b. "Compare countries with annual data from official sources." *The Global Economy.* https://www.theglobaleconomy.com/compare-countries/

The Global Economy. 2020c. "Employment in Agriculture in OECD." *The Global Economy.* https://www.theglobaleconomy.com/rankings/Employment_in_agriculture/OECD/

Thomas, Louise. n. d. *Insect Invaders and the Seduction of Scent.* Wellington, NZ: The Royal Society of New Zealand. file:///Users/mval900/Downloads/Learning+Task+3+-+Insect+Invaders-2.pdf

Timmins, Susan. 1988. *Science and Research Internal Report No.6. Biological Control in Protected Natural Areas.* Wellington, NZ: Department of Conservation. https://www.doc.govt.nz/globalassets/documents/science-and-technical/ir06.pdf

Warburton, Bruce, Phil Cowan, and J. Shepherd. 2009. "How Many Possums Are Now in New Zealand Following Control and How Many Would There Be Without It?" Prepared for Northland Regional Council. *Landcare Research – Manaaki Whenua.* https://envirolink.govt.nz/assets/Envirolink/720-NLRC104-Possum-numbers-inNZ.pdf

Walker, James, David Maxwell Suckling, and C. Howard Wearing. 2017. "Past, Present, and Future of Integrated Control of Apple Pests: The New Zealand Experience." *Annual Review of Entomology* 62: 231–248.

Wallerstein, Immanuel. 2011. *The Modern World-System I: Capitalist Agriculture and the Origins of the European World-Economy in the Sixteenth Century.* Berkeley: University of California Press.

Watts, Meriel. 1994. "The Poisoning of New Zealand." Auckland, NZ: Auckland Institute of Technology Press.

Watts, Jonathan. 2017. "The Amazon effect: how deforestation is starving São Paulo of water." *The Guardian,* November 28. https://www.theguardian.com/cities/2017/nov/28/sao-paulo-water-amazon-deforestation

Wilson, John. 2006. "New Zealand Sovereignty: 1857, 1907, 1947, or 1987?" *New Zealand Parliamentary Library.* https://www.parliament.nz/en/pb/research-papers/document/00PLLawRP07041/new-zealand-sovereignty-1857-1907-1947-or-1987

Wilson, Kerry-Jayne. 2015. "Land birds – overview – Extinct, endangered and threatened species." *Te Ara – the Encyclopedia of New Zealand.* https://teara.govt.nz/en/land-birds-overview/page-4

Wynn, Graeme. 2013. "Destruction under the Guise of Improvement? The Forest, 1840–1920." In *Making a New Land: Environmental Histories of New Zealand,* edited by Eric Pawson and Tom Brooking, 122–138. Dunedin, NZ: Otago University Press.

4 Contextualizing the Aerial Pesticide Spraying Response

When a government decides to eradicate a species there are numerous actions it can pursue, including, but not limited to, destroying host plants, vegetation controls (to prevent residents from accidentally transporting infested plant matter to other regions), biological control (importing a predator or parasite to the species in question), mating disruption technology (distributing sterilized moths), using pheromones to attract male moths, ground spraying of pesticides, and aerial pesticide spraying. In order to better understand why a government agency would pursue an aerial pesticide spraying campaign that would repeatedly place hundreds of thousands of people in harm's way (some multiple times a day and over fifty times over the course of the operation (Office of the Ombudsman 2007), there are a few questions that need to be examined: (1) why would biosecurity authorities turn to pesticides as their initial response?; (2) why did they feel compelled to escalate the operation to include aerial spraying?; and (3) why did they persist with the use of aerial spraying even though they knew it would put tens of thousands of people in harm's way?

The next section tackles the first question by examining the larger cultural context within which this operation took place, elucidating the "synthetic age" (Foster 1999) that industrial societies live in, and how it predisposes people in these societies to address ecosystem problems with synthetic pesticides. While this analysis helps explain MAF's culturally conditioned response to use pesticides at the start of the PAM incursion, it does not explain why MAF felt compelled, more than 30 months into the operation, to escalate its response with aerial pesticide spraying over suburban neighborhoods. To account for that development, the second section examines MAF's mishandling of the initial PAM incursion, which allowed the PAM moth to spread to the point it became a major concern to the forestry industry, who then pressured MAF to escalate its response to PAM. The last section tackles why the government would pursue an aerial spraying operation over suburban neighborhoods despite putting tens of thousands, and eventually hundreds of thousands, in harm's way. Towards that end, I situate the case in its national context, elucidating how New Zealand governments have, historically, demonstrated a consistent and disturbing tendency to put people in harm's way to maximize industry profitability.

DOI: 10.4324/9780429426414-5

Embedding New Zealand's Initial Response in Its Cultural Context

In trying to account for the PAM pesticide spraying operation, it is useful to start by embedding this event within its larger cultural context. As articulated in the last chapter, part of the cultural context is the worldview found in industrial societies, where nature is seen as something to be molded to serve human needs and desires and where the only species that are tolerated are those that do not interfere with the pursuit of those needs and desires. Another important piece of this cultural context is the means people typically use to achieve these environmental objectives, which, in industrialized societies, tends towards the use of synthetic pesticides.

John Bellamy Foster (1999) refers to our era as the "synthetic age," an era where natural products and processes have been increasingly displaced by synthetic products. While one example is the replacement of traditional pest-management techniques (such as crop rotation, companion planting, and ensuring ecosystem biodiversity) with synthetic pesticides, other examples include replacing natural fibers (cotton, wool, silk etc) with synthetic ones and replacing soaps with high-phosphate detergents (*Ibid.*).

Foster (1999) traces this synthetic age to the scientific-technological revolution that occurred in the late nineteenth century, where industry increasingly captured the production of scientific knowledge and reoriented it to better serve their interests. Foster argues this began with the establishment of corporate research labs, which began in the United States when Thomas Edison established his research lab in 1876 and expanded rapidly in subsequent decades, with 300 labs established by 1920 and 2,200 by 1940 (*Ibid.*). These labs increasingly lured researchers from universities and oriented them towards producing knowledge that would better benefit their corporate sponsors. With the increase in industry-sponsored science came a flurry of technological developments aimed at furthering industrialization and benefiting industries. This, in turn, ushered in the chemical revolution of the 1950s, where an expanding array of poisons was developed to eliminate unwanted species.

New Zealand's Embrace of Pesticides

Despite the "Clean and Green" marketing campaign that New Zealand governments have been running over the last two decades, the country has, historically, embraced synthetic pesticides as a means of resolving ecological problems. One indicator is the country's long history of pesticide use, which began in the 1940s, when farmers increasingly turned to DDT, expanded with the arrival of other persistent organochlorine pesticides (such as lindane, aldrin, dieldrin, and chlordane), and, in subsequent decades, expanded further with the arrival of organophosphates, carbamates, and other classes of synthetic pesticide (Manktelow et al. 2005; Watts 1994).

Another indicator of the embrace is the wide variety of pesticides New Zealand governments have approved. For instance, by the late 1980s there were 280–300 pesticide active ingredients registered for use in New Zealand, with an estimated 700–900 pesticide formulations being used (MacIntyre et al. 1989; Watts 1994).

The embrace is also indicated by growing use over time. While industry tends to conceal data about pesticide use, it is estimated that by the late 1980s New Zealand farmers were applying 3,500 to 4,500 tons of pesticide active ingredients per year (Watts 1994). Moreover, while pesticide use is estimated to have declined to slightly less than 3,000 tons by 1999, the following year usage began growing again, as indicated by the 3,500 tons sold in 2000 and the 4,000 tons sold in 2002 (Manktelow et al. 2005). Importantly, these totals do not include the amount used by the timber industry, nor the amounts used by government authorities for controlling pests, such as weeds, possums, and rabbits (Manktelow et al. 2005; Watts 1994). The latter can be significant, as underscored by the fact that, in the mid-1990s, New Zealand authorities were applying approximately 3.7 tons of 1,080 per year to control possums and rabbits, an amount that represented 90% of the world's 1,080 use (Foronda et al. 2007; Peterson et al. 1994; Watts 1994).

A fourth sign of the country's attachment to pesticides is that several governments subsidized the use of these toxic substances, which they did through a variety of schemes between 1970 and 1985: (1) the 1976 Livestock and Incentive Scheme; (2) the 1978 Land Development Encouragement Loan; and (3) the 1981 to 1985 Noxious Plants Control scheme, which reimbursed farmers half of all expenditures on pesticides (Coster et al. 1986).

One product they subsidized was 2,4,5-T, which was one of two chemicals in Agent Orange and was laced with the significant levels of the carcinogenic dioxin TCDD (Coster et al. 1986; Takoko & Gibbs 2004). During the 1970s the New Zealand government actively encouraged the use of 2,4,5-T by subsidizing 116 products that contained it (Takoko & Gibbs 2004). Moreover, between 1981 and 1985 the New Zealand government subsidized this pesticide to the cost of $25 million NZD (Coster et al. 1986: 35). As a result, production and sales grew precipitously in the first years of the subsidy (see Figure 4.1) for the Ivon-Watkins Dow chemical company, which was the herbicide's sole manufacturer in New Zealand.

The country's affinity with pesticides is also signaled by the government's failure to consistently monitor pesticide contamination and its environmental effects. For instance, while the government has consistently monitored groundwater for pesticide contamination, they have failed to adequately monitor pesticide contamination of surface waters or the harm done to ecosystems relying on those surface waters (Ministry for the Environment 2007; Ministry for the Environment & Stats NZ 2019; Taylor et al. 1997). Case-in-point, the contamination of the country's streams was neglected until 2019, at which point researchers found that two or more pesticides were present at 78% of sampled streams and that three or more pesticides were found at 69%

Figure 4.1 NZ Production of 2,4,5-T from 1948 to 1985.
Source: Coster et al. (1986).

of sampled sites (Hageman et al. 2019). If governments do not monitor for such impacts, it removes a disincentive for farmers to use pesticides, which, in turn, encourages their use.

The Government's Reluctance to Ban Harmful Pesticides

The country's embrace of pesticides has been underscored by the government's disturbing reluctance to prohibit pesticides that have been banned in other countries, which has been the case with DDT, chlorpyrifos, methyl bromide, diazinon, atrazine, glyphosate, 2,4-D, dimethoate, 1080, methamidophos, 2,4,5-T, and numerous others (Castle 2020; Johnsen 2018; Wall 2018).

In the case of DDT, while it was banned in Sweden, Norway, West Germany, Hungary, and the United States by 1972, New Zealand did not ban it until 1989, a full 17 years later (Hayes 1969; Mandavilli 2006; Taylor et al. 1997). Reflecting this fact is that by 2000 DDT breakdown products were still appearing far more often in New Zealand dairy products than in dairy products from the United States (68% vs. 3.4% of samples) (White 2000).

Another example is chlorpyrifos, the chemical MAF used at the start of the PAM incursion, which can cause numerous medical problems, including: neurodevelopmental problems, pervasive development disorder, symptoms related to attention-deficit hyperactivity disorder, psychiatric disturbances, immune suppression, low-birth weight, dizziness, fatigue, runny nose or eyes, salivation, nausea, intestinal discomfort, sweating, changes in heart rate, paralysis, seizures, loss of consciousness, and death (ATSDR 1998; CHE 2020a; Rauh et al. 2011). In 2005 the United States banned it for home garden and non-agricultural use (Collins & Johnston 2005) and several European countries have instituted total bans (including Denmark, Finland, Germany, Sweden, Switzerland, and the United Kingdom) (Dahllöf 2019). And yet, it is still widely used in New Zealand agriculture, as underscored by the fact it was recently detected in 83% of New Zealand streams (Balance 2016; Hageman et al. 2019).

A third example is diazinon, an organophosphate pesticide that can cause nausea, vomiting, abdominal cramps, diarrhea, breathing difficulties, damaged sperm, premature births, headaches, dizziness, weakness, impaired eyesight, passing out, and coma (ATSDR 2015; CHE 2020b). This pesticide is also particularly dangerous to birds, and in the 1990s it was responsible for more bird deaths in the United States than any other pesticide (Cone 2005). Subsequently, the United States banned it from residential use in 2005, and in 2006 the European Union banned all its uses (Cone 2005; PAN Europe 2006). In New Zealand, however, a country that prides itself on its bird life and which markets itself as "100% Pure and Natural," the chemical is still sold over the counter and will not be banned until at least July 1, 2028 (Environmental Protection Authority 2013).

Still another disturbing case is the government's handling of 2,4,5-T, a chemical made infamous by the fact it was one of two chemicals used to create Agent Orange, the potent herbicide the US military used to defoliate forests and destroy crops during the Vietnam War (Institute of Medicine 1994). 2,4,5-T is associated with numerous health problems, including chloracne, peripheral neuropathy, headaches, digestive disorders, muscular weakness, animal and human birth defects, depression, porphyria, and soft tissue sarcoma (CHE 2020c; Takoko & Gibbs 2004; Watts 1994; Wildblood-Crawford 2008). Moreover, it is laced with the dioxin 2,3,7,8-tetrachlorodibenzo-p-dioxin (TCDD), which is known to be a potent human carcinogen (NIH/NIEHS 2001). Despite these significant health effects, New Zealand was late to move away from 2,4,5-T as well. While Cyprus, Sweden, India, Japan, Guatemala, Norway, the Netherlands, Turkey, and the United States had all banned the chemical by 1979, between 1981 and 1985 the New Zealand was encouraging its use through subsidies. Moreover, the country did not move on from that herbicide until 1989, at which point it was the last country to do so (Coster et al. 1986; United Nations Environmental Programme 1996).

A History of Using Pesticides for National Control Operations

Besides the significant use for agricultural purposes, the country also has a long history of using synthetic chemicals for large-scale pest control programs, including those aimed at controlling plant species (like gorse), mammals (including rabbits, possums, and stouts), and invertebrates (such as moths and subterranean termites) (Office of the Ombudsman 2007; PCE 2000). This practice dates back to at least 1954, when New Zealand began importing

1080 to control the rabbit population (Rammell & Fleming 1978). Moreover, over time government agencies began using this pesticide to also control possums, with an estimated 3.7 tons of 1080 being used in 1993 alone, at a cost of $27 million (Peterson et al. 1994; Watts 1994; Weaver 2003). Other large-scale pest control operations have included those aimed at controlling weeds on roadsides and in waterways, both of which have faced noticeable opposition in urban areas (Watts 1994).

Beyond administering pesticides through ground-based traps, biosecurity agencies have also taken to administering them by aerial spraying, which they have been doing with 1080 for decades (Weaver 2003). As well, in 1996 and 1997 the New Zealand government carried out a large-scale aerial pesticide spraying operation in East Auckland, in order to eliminate the white-spotted tussock moth (WSTM), which, prior to the PAM campaign, was the largest aerial pesticide spraying campaign carried out in an urban area (Office of the Ombudsman 2007).

Given the country's general affinity for using synthetic pesticides, the government encouraging their use in multiple ways, and the government's long history of using such pesticides for national pest control programs, it is not surprising that MAF's initial response to PAM was to use pesticides. However, this does not explain why they went from ground spraying pesticides in a small and mostly industrial area to carrying out the largest urban aerial pesticide spraying operation in world history.

In order to understand why MAF escalated to using aerial pesticide spraying, it is necessary to understand the institutional context within which that decision took place. This includes considering the preceding 1996–1997 WSTM eradication operation that also relied on aerial spraying, a critical government report about that eradication operation, and the numerous mistakes MAF leadership made in its initial response to the PAM incursion.

Project Ever Green and MAF's Ineffective Initial Response to PAM

To properly understand the PAM case we need to consider Operation Ever Green, which was New Zealand's 1996–1997 operation to eradicate the WSTM from East Auckland. Project Ever Green had many parallels with the PAM eradication operation. Like PAM, the WSTM was viewed as a threat to the country's primary industries, which triggered an eradication response (Hosking et al. 2003). Other similarities include that biosecurity officials struggled to make progress against the moths, suffered through periods where they could not keep the public on side, and eventually turned to aerial pesticide spraying (Panckhurst 2001). Besides the parallels between the two operations, Operation Ever Green was critical for the PAM operation, as it established a New Zealand precedent for pursuing an extensive aerial pesticide spraying operation over suburban areas. As well, it demonstrated that such an operation could eradicate an invasive species.

The public's response to Operation Ever Green was mixed. On the one hand, the aerial spraying generated significant opposition among East Auckland residents (Office of the Ombudsman 2007). On the other hand, there was a certain amount of tolerance because the public understood that this was the first time biosecurity officials were attempting such an operation and that the kinks needed to be worked out. Moreover, while the Ever Green team struggled to figure out the logistics of the ground-breaking operation, there was positive sentiment about the collegiality and cohesiveness experienced in that group, which, many believed, positioned New Zealand Biosecurity to effectively handle future incursions (Panckhurst 2001). Moreover, in light of the subsequent PAM operation, activists came to recognize and appreciate the higher level of transparency that government officials offered during Operation Ever Green, as well as their willingness to work constructively with the local community (personal communication with PAM activist).

Striking a more discordant note was the government-ordered review of Operation Ever Green's cost and effectiveness (Panckhurst 2001). Among other things, the reviewers criticized the operation for poor decision-making, unclear goals, failing to sufficiently rely on science, and the project's $13 million price tag, which was inflated by the need to use airplanes to administer the pesticide (*Ibid.*). The review was itself criticized for being at odds with general sentiment about the operation and for making what many key players considered to be unfair criticisms (OAG 2002; Pankhurst 2001).

In December 1998 a significant and fateful change in leadership occurred as Ruth Frampton was named to replace Gordon Hoskins (who oversaw Operation Ever Green) as MAF's Chief Technical Officer (Forestry), a position that was later retitled to Director of Forest Biosecurity. Frampton's credentials included being a MAF national advisor on plant pest surveillance, a former biosecurity advisor, running a successful operation to eradicate the Mediterranean fruit fly, and being one of the three authors of the maligned government-ordered review of the Operation Ever Green (Pankhurst 2001).

Her appointment to this position was fateful to the PAM operation for two reasons. First, it antagonized key players in forest biosecurity, who felt she and her colleagues unfairly criticized Operation Ever Green in their report (Pankhurst 2001). In turn, such antagonisms interfered with her ability to effectively recruit and work with those players to eradicate the painted apple moth. Second, her management of the PAM operation was marked by a series of missteps, which allowed the PAM population to spread and which, effectively, turned the incursion into an infestation that increased the likelihood that aerial pesticide spraying would be called for.

One misstep was significantly delaying the breeding of research moths. Breeding a research population is crucial as it enables other research that is essential to effectively manage an incursion, including: (1) carrying out host feeding trials (to see which local flora the invasive species is likely to feed on); (2) assessing the spread of the moth (by using trapped live females to lure males); (3) evaluating life cycles; (4) developing synthetic pheromones (to replace live females as bait); and (5) conducting pesticide efficacy trials

(Hosking et al. 2003; PAM Community Advisory Group 2001; OAG 2002; Panckhurst 2001). Three months elapsed before Frampton appointed a lab to develop a breeding colony, which delayed the start of the other research tasks (OAG 2002).

Another misstep was contracting the moth breeding work to Hort Research labs, instead of Forestry Research Institute, in Rotorua, which had played a significant role in breeding the research colony for the WSTM eradication, had just opened a state-of-the-art quarantine facility to carry out such work, and had offered to breed a research colony of painted apple moths for free (OAG 2002). Her choice ended up backfiring when the two research colonies bred by Hort Research died prematurely, which brought other eradication tasks (including host feeding trials, life-cycle assessment, pheromone development, and pesticide testing) to a halt. The problem was highlighted by the Auditor-General (2002), who tied breeding delays to the agency's failure to eradicate the moth in a timely manner: "had MAF contracted for rearing earlier or had MAF utilized the rearing facility in Forest Research, Rotorua these problems would have been solved earlier and progress on PAM eradication would be more advanced at this point" (p. 65). Moreover, he chastised the director for not accepting Forestry Labs' offer to breed the moths for free, emphasizing that, at a minimum, Forestry Labs should have independently bred a research colony as a backup to the Hort Labs effort. The problem was also highlighted by the independent reviewers to MAF's response, who noted Hort Research Labs' ineffectiveness during the first two years of the operation and specifically recommended that MAF contract the moth breeding work to Forestry Research Institute (*Ibid.*).

A third misstep occurred around the development of a synthetic pheromone, which is essential to strengthen monitoring work and reduce reliance on a breeding population. Shortly after PAM's discovery, in April 1999, the entomologists who had successfully developed the synthetic pheromone for the WSTM (i.e. John Clearwater and Gerhard Gries) contacted Frampton to offer their services at a deep discount compared to the Hort Research bid (Goven et al. 2007; OAG 2002). Not only did the operation director reject the offer from world experts with a strong track record in the New Zealand context, she failed to even extend the courtesy of responding to their offer (*Ibid.*). Moreover, her refusal to contact them persisted even when other pheromone development efforts floundered and when independent reviewers, more than two years after the moth's discovery, specifically recommended that MAF hire the Clearwater and Gries team (*Ibid.*). The Auditor-General (2002) viewed this mistake as also contributing to MAF's failed initial response:

> Failure to establish a breeding colony of PAMs at Forest Research in Roturua, and to properly consider an expert's offer to develop a synthetic pheromone for use in the response, resulted in delays that have led to the PAM spreading from the original infestation site.
>
> (p. 67)

The consultants were not contacted until right before the September 2001 Technical Advisory Group meeting, at which point the government was already preparing for an aerial pesticide spraying operation (*Ibid.*). While MAF eventually contacted the experts, the Auditor-General (2002) underscored that contacting them at such a late date significantly hampered eradication efforts: "MAF could have taken this course of action some two and a half years earlier, and there is a strong likelihood that the response to the PAM would have been different if it had done so" (p. 69).

Another blunder pertained to MAF's initial ground-spraying activities, which consisted of using chlorpyrifos and deltamethrin, both of which are environmentally damaging and toxic to humans (PAM CAM 2001). Deltamethrin, in particular, is quite toxic to the marine environment, and regulations prevented its application near the banks of streams, rivers, and other bodies of water, where PAM was able to refuge itself and multiply (personal communication with scientist). For several months the PAM Community Advisory Group (2001) requested that MAF pursue ground spraying with Btk instead of the more toxic deltamethrin, but MAF refused to acknowledge advisory group's suggestion, let alone transition to the less toxic Btk. The PAM Community Advisory Group (2001) maintains that if MAF had used the less toxic Btk in the initial ground-spraying operations instead of the more toxic pesticides, the moth population would have never been able to multiply the way it did, as biosecurity agents would have been able to more thoroughly spray along the harbor and creek edges, which would have eliminated the main infestations.

Another management failure was the Director's unwillingness to provide a clear operational plan, which was first noted in November 1999, when MAF's Forest Biosecurity Advisory Committee highlighted there was no operational plan, even though seven months had elapsed since the moth's discovery (OAG 2002). The Director informed the Auditor-General (2002) that "such plans were of limited use in responses to biological organisms because of the complexity and unpredictability of a response" (p. 70). However, the Auditor-General (2002) disagreed, stating the complexity and unpredictability of responses actually reinforces the importance of having detailed plans to be able to (1) "prepare for unforeseen events"; (2) "ensure that action is effectively co-ordinated"; and (3) be used as a basis for communicating with stakeholders about the action being taken" (p. 70). By May 2001, two years after the moth's initial incursion, the problem was still present as Cabinet had yet to receive an operational plan for eradicating the moth (Panckhurst 2001). This was a source of concern for the Department of Conservation, the Ministry of the Environment, and the Ministry of Research, Science and Technology, who were described as "concerned that two years after the moth's detection in Auckland, no operational plan to eradicate the moth has been prepared, approved and implemented" (Cabinet Infrastructure and Environment 2001: 8, as cited by Panckhurst 2001). It was not until November 30, 2001, that a plan was produced, which was 31 months after the moth was first discovered (OAG 2002).

The Director's numerous missteps enabled the PAM population to grow exponentially over time, which transformed a limited incursion into a full-blown infestation. As one entomologist quipped "if this was an attempt to introduce a new species, you'd have to be pleased at how it was going" (Peter Maddison, as cited in Panckhurst 2001: 58). Moreover, in September 2001 Gordon Hosking (Director of the 1996–1997 operation to eradicate WSTM), stated that, because of PAM's spread, it would be very difficult to eradicate it and that he could not imagine doing so without an aerial spraying operation that was larger than the one used to eradicate WSTM and "perhaps bigger than anyone could countenance" (as cited in Panckhurst 2001: 58).

To bring the mismanagement into sharper relief, it is useful to compare the PAM eradication operation to the 1996–1997 operation to eradicate the WSTM. The latter was able to be completed in less time (7 months vs 29 months), with fewer sprayings (23 vs over 45), with a smaller spray zone [4,000 hectares (10,000 acres) vs 10,600 hectares (26,913 acres)], and at a lower cost ($13 million vs $65 million) than its PAM counterpart (Goven et al. 2007; Office of the Ombudsman 2007; Swinney 2004). Moreover, the WSTM eradication team was able to accomplish this despite having a more challenging eradication to carry out: (1) whereas PAM only occupied 5 hectares (12 acres) at the start of the eradication operation, WSTM was established on 700 hectares (1,729 acres) at the start of its eradication operation; (2) the female painted apple moth was wingless and thus had reduced capacity to spread the infestation, whereas this was not the case with the WSTM; (3) where PAM of all ages died when exposed to the Btk pesticide, the largest WSTM caterpillars survived; and (4) the WSTM eradication did not have a prior case to draw on, whereas the PAM eradication team could draw on the experiences from the WSTM eradication (Swinney 2004).

MAF's inability to contain, let alone eradicate PAM, raised concerns among plantation forestry officials, who believed that if the moth reached forest plantations, some of the trees would experience light defoliation, which would slow tree growth and prevent profit maximization. Industry concerns were first communicated to the MAF Group Director of Biosecurity in November 1999 and continued to be communicated as the infestation spread (OAG 2002). Moreover, in 2001 industry representatives began pressing for an independent review of the Director of Forest Biosecurity's handling of the PAM incursion, while also pressing for aerial pesticide spraying of the incursion sites. In the first quarter of 2001 the MAF Group Director of Biosecurity acquiesced to the pressure by commissioning an independent review of MAF's response to PAM, which was published in June 2001 and made 34 recommendations, including proceeding with an aerial pesticide spraying operation (*Ibid.*).

In turn, MAF began planning a targeted aerial pesticide spraying operation that would use helicopters to spray infestation hotspots and attain difficult-to-reach areas. Although MAF began implementing the plan in January 2002, in February 2002 the Director of Forest Biosecurity signaled the agency would consider pulling away from the pesticide spraying if it did not achieve the expected results, at which point they would have to consider controlling the pest rather than eliminating it (Beston 2002). This drew a sharp rebuke from the Forest Industries Council, whose chief executive (i.e. Rob McLagan) stated that "opting out of eradication was not acceptable" (as cited in Beston 2002). As we now know, even though the targeted spraying failed to achieve the desired results (i.e. eradication), MAF doggedly doubled down on the strategy, increasing the number of sprayings from a maximum of 8 to over 40 and dramatically expanding the spray zone from 500 hectares (1,235 acres) to over 10,600 hectares (26,913 acres) (Goven et al. 2007; Office of the Ombudsman 2007).

An irony of the expansion is that it could and should have been unnecessary. When she was appointed to be the Director of MAF Forest Biosecurity Ruth Frampton inherited the infrastructure and team that had proven effective at eliminating the WSTM incursion. Additionally, she had the opportunity to benefit from the experience and knowledge MAF gained during Operation Ever Green. Consequently, when PAM established a toehold in Auckland, Frampton should have been able to act swiftly, decisively, and effectively to eliminate it. Instead, however, she allowed personal grudges to marginalize key players, which undermined the agency's capacity to quickly and effectively eradicate PAM (Goven et al. 2007; OAG 2002; PAM Community Advisory Group 2001).

Another irony is that the criticisms she and her co-authors made of Project Ever Green, in the government-ordered review, were even more glaringly present in her handling of the PAM operation. For instance, while she critiqued Operation Ever Green for its lack of clear direction, her entire management of the PAM operation was marked by a lack of clear planning, as underscored by the fact her team did not produce an operational plan until the 31st month of the incursion (OAG 2002; PAM Community Advisory Group 2001). Moreover, while she criticized Operation Ever Green for resorting to aerial pesticide spraying, her numerous managerial missteps all but guaranteed that an even larger aerial pesticide spraying operation would be used to eradicate PAM. Lastly, while she critiqued the $13 million price tag associated with Operation Ever Green, her inept handling of the PAM operation contributed significantly to its final price tag of over $65 million (Office of the Ombudsman 2007), which did not include the cost of long-term health effects (including physical and psychological) for the 200,000 residents who were repeatedly exposed to the spraying, nor productivity losses associated with those effects.

MAF's missteps help explain how a relatively straightforward incursion turned into a full-blown infestation, which, in turn, created pressure for a much more extensive and aggressive response. Illuminating MAF's missteps helps us better understand how the agency created a situation that forced it to escalate its response to the moth and to consider using aerial pesticide spraying as part of that escalated response. However, what remains unexplained is why government officials pursued this option even though it would place 13,000 people, and eventually hundreds of thousands, in harm's way.

Government ignorance is not an adequate explanation because when government officials were planning the PAM spraying operation they were well aware of health issues related to the Foray 48B pesticide. This included the health impacts related to the 1996–1997 Project Evergreen pesticide spraying operation, which led 350 residents to report illness symptoms related to the spraying (Watts 2003). Additionally, government officials were well aware of the 2003 Blackmore report, which identified that nearly 400 residents had suffered health effects from the first 12 months of the PAM spraying operation. This evidence seemingly did little to curtail spraying operations, which continued for an additional 17 months after Hana Blackmore released her report.

The Proclivity for Sacrificing Public Health on the Altar of Economic Growth

To better understand why government officials would readily place tens of thousands of people in harm's way, we need to situate the decision within the larger political economy.

Towards that end, it is useful to draw on the work of social scientists who have shown that government agencies have a tendency to make decisions that prioritize industry interests over those of citizens. For instance, Hans Baer's (1990) work demonstrates how United States government officials have consistently protected the nuclear industry at the expense of protecting citizens. Similarly, Daniel Faber (2008) shows that government officials have supported the pollution-industrial complex over citizen health. These examples reflect David Pellow's (2015) point that Western states do not equally serve a plurality of interests but rather tend to prioritize the interests of an economic elite. Complementing Pellow's perspective is Schnaiberg and Gould's (1994) "treadmill of production" framework, which posits that the driver of government actions is their "economic growth at all cost" mindset. Under this mindset, government agencies pursue actions geared towards helping industries maximize profits (such as providing subsidies and failing to properly enforce regulations), at the expense of people and the environment. This framework is quite germane for explaining the New Zealand case, whose governments have had a long history of supporting or even perpetrating actions that placed people in harm's way, in order to maximize industry profitability.

A prime example are their actions vis-à-vis pesticides, which included the previously discussed willingness to approve a wide assortment of harmful pesticides, as well as a willingness to encourage pesticide use via financial subsidies. Additionally, the pattern is strongly exemplified by the government's consistent reluctance to prohibit pesticides that have been shown to be harmful and that have been banned in other countries, as discussed above.

The government's tendency to place people in harm's way has also manifested itself for other harmful chemicals. For example, for over 40 years New Zealand governments allowed sawmills to use pentachlorophenol (PCP) to treat timber, creating over 7,500 sites of contamination throughout the country (National Task Group 1992). PCP is associated with a host of health problems, including nausea, fatigue, fever, acute tubular necrosis, immune suppression, pancreatic cancer, renal cancer, aplastic anemia, chloracne, cognitive impairment, mood swings, and developmental delay (CHE 2020d; Dew 1999). Moreover, commercial grades used in New Zealand were particularly harmful because they were laced with dioxin and furan, both of which can cause reproductive abnormalities, immunological disorders, and cancer, even in small doses (Szabo 1993). While the government eventually de-registered the chemical in 1991, it only did so after allowing thousands of workers to be exposed to the chemical. Moreover, the de-registering occurred after allowing significant contamination of numerous environmental locations, including Lake Rotorua, a popular tourist destination whose dioxin level in the early 1990s was twice as high as Lake Ontario (*Ibid.*). A similar situation occurred at Hanmer Springs, a popular New Zealand holiday campground that had soil, dust, and groundwater PCP levels that were beyond what would trigger remedial action in Canada and the Netherlands (*Ibid.*).

The propensity to put citizens in harm's way has also manifested itself with the disposal of chemicals. For instance, from the mid-1970s until the late 1980s the New Zealand government allowed the Ivon-Watkins Dow chemical manufacturer to incinerate dioxin-contaminated production wastes near a residential neighborhood (Wildblood-Crawford 2008). Notably, this persisted throughout the late 1970s, even though the 1976 Seveso (Italy) accident had focused worldwide attention on the harms associated with dioxins (De Marchi 1997).

Additionally, in the 1970s the government authorized the same chemical manufacturer to bury 632 drums of chemical waste (which included 2,4-D and 2,4,5-T) in unlined pits, which began leaking into the river and sea in the mid-1980s (The Daily News 1985, 1988; Tonkin & Taylor Limited 2006). Relatedly, in 2009 storms unearthed two crushed drums of chemical waste in a local public park, which had previously existed as a landfill (Lidgard & Profitt 2009; Ministry for the Environment 2011). The cleanup of Marfell Park cost taxpayers more than $180,000, and the government refused to hold the company accountable. Their rationale was that when the company buried the drums in the 1970s, the government had given them permission to do so (Dominion Post 2009).

What these examples illustrate is that the New Zealand government has demonstrated a disturbing tendency to allow and even encourage activities that place citizens in harm's way. This suggests that, similar to David Pellow's (2015) analysis about the United States, New Zealand governments are not as pluralistic as one would like and are instead systematically oriented towards benefitting the economic elite. Moreover, the examples correspond with Schnaiberg and Gould's (1994) treadmill of production framework, which suggests that nation states are fixated with maximizing growth at all costs. To be fair, New Zealand is far from being the only such case, as social scientists have revealed many other cases where nation states place industry profits ahead of human health. However, the New Zealand government's ecological violence is particularly jarring when we juxtapose it to its annual greenwashing marketing campaign to portray the country as "Clean and Green" and "100% Pure and Natural." As this relates to the PAM case, when we consider that the forestry industry was concerned about how the moth *might* impede their ability to maximize profits, it should come as no surprise that government officials willingly placed tens of thousands of citizens in harm's way and continued doing so even when the number expanded to hundreds of thousands.

Summary

This chapter's overarching objective was to illuminate key factors that contributed to MAF's decision to pursue an aerial pesticide spraying operation to eliminate the painted apple moth. Towards that end, this chapter related the case to the dominant cultural norms in industrialized societies, where the use of synthetic pesticides has become the culturally conditioned response to ecosystem problems. This chapter also related the pursuit of aerial spraying to the specifics of the case, which included the government's proclivity for pesticides and MAF's mismanagement of the initial PAM incursion. This mismanagement enabled the moth population to spread significantly and created the perceived need for an escalated response. As well, this chapter emphasized that pursuing an aerial spraying operation over urban areas was enabled by a political economy that prioritizes maximizing industry profits over the health and well-being of people.

Although this chapter illuminates factors that compelled the New Zealand government to pursue the aerial pesticide spraying campaign, we have yet to consider how the local community responded to the government's plan. The citizens' response is important because pesticide use is underpinned by public acceptance. Without public acceptance, consumers will be more reluctant to use pesticides in their gardens and citizens will be more likely to pressure for tougher regulations and bans. An example is the province of Québec (Canada), where citizen pressure led Montréal city officials to ban glyphosate at the end of 2019 (Montreal Gazette 2019) and provincial officials to issue a province-wide ban of atrazine, chlorpyrifos, and three neonicotinoids (clothianidin, imidacloprid, and thiamethoxam) in 2018 (Fletcher 2018). While public acceptance is important for the use of consumer pesticide products, it is particularly important for aerial pesticide spraying operations in urban environments, as exposing large numbers of people to pesticides can trigger significant political backlash that will derail the government's plan, as occurred with the 2008–2010 Light Brown Apple Moth eradication campaign in Northern California (Spiegelman 2010).

To shed more light on this issue, the next chapter discusses the community's opposition to the government plan. Moreover, it begins the longer conversation about why that opposition failed to stop the spraying operation, which includes the government's exploitation of a regulatory loophole and the slow spread of opposition among the masses.

References

Agency for Toxic Substances and Disease Registry (ATSDR). 1998. "Public Health Statement for Chlorpyrifos." Toxic Substances Portal, U.S. Government. https://www.atsdr.cdc.gov/ToxProfiles/tp84-c1-b.pdf

Agency for Toxic Substances and Disease Registry (ATSDR). 2015. "Public Health Statement for Diazinon." Toxic Substances Portal, U.S. Government. https://wwwn.cdc.gov/TSP/PHS/PHS.aspx?phsid=511&toxid=90

Baer, Hans. 1990. "Kerr-McGee and the NRC: From Indian Country to Silkwood to Gore." *Social Science & Medicine* 30(2): 237–248.

Balance, Alison. 2016. "Pesticide bad news for bee learning and memory." *Radio New Zealand.* https://www.rnz.co.nz/national/programmes/ourchangingworld/audio/201804427/pesticide-bad-news-for-bee-learning-and-memory

Beston, Anne. 2002. "$11m moth blitz risking failure." *New Zealand Herald*, February 15. https://www.nzherald.co.nz/nz/news/article.cfm?c_id=1&objectid=939414

Castle, Belinda. 2020. "Pesticides in fruit and vege." *Consumer.org.* https://www.consumer.org.nz/articles/pesticides-in-fruit-and-vege

Collaboration on Health and the Environment (CHE). 2020a. *Toxicants and Disease Database: Chlorpyrifos.* https://www.healthandenvironment.org/our-work/toxicant-and-disease-database/?showcategory=&showdisease=&showcontaminant=2696&showcas=&showkeyword=

Collaboration on Health and the Environment (CHE). 2020b. *Toxicants and Disease Database: Diazinon.* https://www.healthandenvironment.org/our-work/toxicant-and-disease-database/?showcategory=&showdisease=&showcontaminant=2485&showcas=&showkeyword=

Collaboration on Health and the Environment (CHE). 2020c. *Toxicants and Disease Database: 2,4,5-T.* https://www.healthandenvironment.org/our-work/toxicant-and-disease-database/?showcategory=&showdisease=&showcontaminant=2488&showcas=&showkeyword=

Collaboration on Health and the Environment (CHE). 2020d. *Toxicants and Disease Database: Pentachlorophenol (PCP).* https://www.healthandenvironment.org/our-work/toxicant-and-disease-database/?showcategory=&showdisease=&showcontaminant=2330&showcas=&showkeyword=

Collins and Johnston. 2005. "NZ using pesticides barred in America." *New Zealand Herald*, January 13, 2005. https://www.nzherald.co.nz/nz/news/article.cfm?c_id=1&objectid=10006400

Cone, Marla. 2005. "EPA Takes Pest Killer Diazinon Off the Shelves." *Los Angeles Times*, January 1. https://www.latimes.com/archives/la-xpm-2005-jan-01-na-pest1-story.html

Coster, A. P., A. S. Edmonds, J. M. Fitzsimons, C. G. Goodrich, J. K. Howard, and J. R. Tustin. 1986. *The Use of 2,4,5-T in New Zealand: A Report to the Environmental Council.* A Report by a Working Party to the Environmental Council. Wellington, New Zealand: Commission for the Environment for the Environmental Council.

Dahllöf, Staffan. 2019. "The most dangerous pesticide you've never heard of." *EU Observer*, June 17 2019. https://euobserver.com/health/145146

De Marchi, Bruna. 1997. "Seveso: From Pollution to Regulation." *International Journal of Environment and Pollution* 7(4). doi:10.1504/IJEP.1997.028318

Dew, Kevin. 1999. "National Identity and Controversy: New Zealand's Clean Green Image and Pentachlorophenol." *Health and Place* 5: 45–57.

Dominion Post. 2009. "Death by Dioxin." *Stuff.co.nz*, January 31, 2009 (accessed August 2, 2021). www.stuff.co.nz/national/health/403166/Death-by-dioxin

Environmental Protection Authority. 2013. "Safely using insecticides containing diazinon on plants." *New Zealand Government*. https://worksafe.govt.nz/search/SearchForm/?Search=diazinon&action_results=Go

Faber, Daniel. 2008. *Capitalizing on Environmental Injustice: The Polluter-Industrial Complex in the Age of Globalization*. Lanham, MD: Rowman and Littlefield Publishers.

Fletcher, Raquel. 2018. "Québec tightens rules on pesticides." *Global News*, February 19.

Foronda, Natalia M., Jefferson Fowles, Nerida Smith, Michael Taylor, and Wayne Temple. 2007. "A Benchmark Dose Analysis for Sodium Monofluoroacetate (1080) Using Dichotomous Toxicity Data." *Regulatory Toxicology and Pharmacology* 47(1): 84–89.

Foster, John Bellamy. 1999. *The Vulnerable Planet*. New York: Monthly Review Press.

Goven, Joanna, Tom Kerns, Romeo Quijano, and Dell Wihongi. 2007. *Report of the March 2006 People's Inquiry Into the Impacts and Effects of Aerial Spraying Pesticide over Urban Areas of Auckland, for the People's Inquiry*.

Hageman, Kimberly, Christopher Aebig, Kim Hoang Luong, Sarit L. Kaserzon, Charles Wong, Tim Reeks, Michelle Greenwood, Samuel Macaulay, and Christoph Matthaei. 2019. "Current-Use Pesticides in New Zealand Streams: Comparing Results from Grab Samples and Three Types of Passive Samplers." *Environmental Pollution* 254. https://doi.org/10.1016/j.envpol.2019.112973

Hayes, Wayland. 1969. "Sweden Bans DDT." *Archives of Environmental Health: An International Journal* 18(6): 872. doi:10.1080/00039896.1969.10665507

Hosking, Gordon, John Clearwater, John Handiside, Malcolm Kay, J. Ray, and N. Simmons. 2003. "Tussock Moth Eradication – A Success Story from New Zealand." *International Journal of Pest Management* 49(1): 17–24.

Institute of Medicine. 1994. *Veterans and Agent Orange: Health Effects of Herbicides Used in Vietnam. Committee to Review the Health Effects in Vietnam Veterans of Exposure to Herbicides*. Washington, DC: National Academy Press.

Johnsen, Meriana. 2018. "Controversial chemicals not on new safety review list." *Radio New Zealand*, October 16, 2018. https://www.rnz.co.nz/news/national/368744/controversial-chemicals-not-on-new-safety-review-list

Lidgard, Rod and Graeme Profitt. 2009. *Marfell Park, New Plymouth, Environmental Investigation*. Prepared for Taranaki Regional Council by Pattle Delamore Partners Ltd, Wellington, NZ.

Mandavilli, Apoorva. 2006. "DDT Returns." *Nature Medicine* 12(8): 870–871.

Manktelow, D., P. Stevens, J. Walker, S. Gurnsey, N. Park, J. Zabkiewicz, D. Teulon, and A. Rahman. 2005. *Trends in pesticide use in New Zealand 2004*. Report to the Ministry for the Environment. Project No. SMF4193.78p

MacIntyre, Angus, Nicholas Allison, and David Penman, Carolyn E O'Fallon and Brian A Croft. 1989. *Pesticides: Issues and Options for New Zealand*. Ministry for the Environment. Wellington, N.Z.: The Ministry.

Ministry for the Environment. 2007. *Environment New Zealand 2007*. https://www.mfe.govt.nz/sites/default/files/environment-nz07-dec07.pdf

Ministry for the Environment. 2011. *New Zealand Inventory of Dioxin Emissions to Air, Land and Water, and Reservoir Sources: 2011.* Wellington, NZ: Ministry for the Environment.

Ministry for the Environment and Stats NZ. 2019. *New Zealand's Environmental Reporting Series: Environment Aotearoa 2019.*

Montreal Gazette. 2019. "Glyphosate: Montreal to ban herbicide used in Roundup." *Montreal Gazette,* September 5.

National Task Group. 1992. *Report of the National Task Group Investigating Site Contamination from the Use of Timber Treatment Chemicals.* Wellington, NZ: Ministry for the Environment.

National Institutes of Health/National Institute Of Environmental Health Sciences (NIH/NIEHS). 2001. "TCDD-Dioxin-Is Listed as "Known Human Carcinogen" in Federal Government's Ninth Report On Carcinogens." *ScienceDaily,* January 23, 2001. www.sciencedaily.com/releases/2001/01/010123074358.htm

Office of the Controller and Auditor-General (OAG). 2002. *Report of the Controller and Auditor-General: Management of Biosecurity Risks: Case Studies.* Wellington, NZ: The Audit Office. https://oag.parliament.nz/2002/biosecurity-case-studies/docs/part3.pdf

Office of the Ombudsman. 2007. *Report of the Opinion of Ombudsman Mel Smith on Complaints Arising from Aerial Spraying of the Biological Insecticide Foray 48B on the Population of Parts of Auckland and Hamilton to Destroy Incursions of Painted Apple Moths, and Asian Gypsy Moths, Respectively.* Wellington, NZ: Office of the Ombudsman.

PAM Community Advisory Group. 2001. "Time to Get Ruth-Less with the Painted Apple Moth." *Scoop Independent News,* December 17. https://www.scoop.co.nz/stories/PO0112/S00077/get-ruth-less-with-the-painted-apple-moth.htm?from-mobile=bottom-link-01

Pesticide Action Network (PAN) Europe. 2006. "What substances are banned and authorised in the EU market?" https://www.pan-europe.info/old/Archive/About%20pesticides/Banned%20and%20authorised.htm

Panckhurst, Paul 2001. "mothbeaten." *Metro Magazine,* October 2001.

Parliamentary Commissioner for the Environment (PCE). 2000. *New Zealand under Siege: A Review of the Management of Biosecurity Risks to the Environment.* Wellington, NZ: Office of the Parliamentary Commissioner for the Environment.

Pellow, David. 2015. "The State and Policy: Imperialism, Exclusion, and Ecological Violence in State Policy." In *Twenty Lessons in Environmental Sociology,* edited by K.A. Gould and T.L. Lewis, 53–66. Oxford, UK: Oxford University Press.

Peterson, Dana R., P. Blaschke, D. Gibbs, B. Gordon, and P. Hughes. 1994. *Possum Management in New Zealand.* Office of the Parliamentary Commissioner for the Environment.

Rammell, Colin George and Peter A. Fleming. 1978. *Compound 1080. Properties and Use of Sodium Monfluoroacetate in New Zealand.* Wellington, NZ: Animal Health Division, Ministry of Agriculture and Fisheries.

Rauh, Virginia, Srikesh Arunajadai, Megan Horton, Frederica Perera, Lori Hoepner, Dana Barr, and Robin Whyatt. 2011. "Seven-Year Neurodevelopmental Scores and Prenatal Exposure to Chlorpyrifos, a Common Agricultural Pesticide." *Environmental Health Perspectives* 119(8): 1196–1201.

Schnaiberg, Allan and Kenneth Alan Gould. 1994. *Environment and Society: The Enduring Conflict.* Caldwell, NJ: Blackburn Press.

Spiegelman, Annie. 2010. "The Light Brown Apple Moth (LBAM) Doesn't Deserve the Starring Role." *Huffington Post*, June 14, 2010. https://www.huffpost.com/entry/the-light-brown-apple-mot_b_523306

Swinney, Clare. 2004. "A Bug-Fuelled Biohazard." *Investigate* July: 56–61.

Szabo, Michael. 1993. "New Zealand's Poisoned Paradise." *New Scientist* 31: 29–33.

Takoko, Mere, and Andrew Gibbs. 2004. *Nga Matitapu o te Hakino: People Poisoned Daily.* Auckland: Greenpeace NZ.

Taylor, Rowan, Ian Smith, Peter Cochrane, Brigit Stephenson, and Nicci Gibbs. 1997. *The State of New Zealand's Environment 1997.* Wellington, NZ: Ministry for the Environment. ISBN 0-478-09000-5.

The Daily News. 1985. "IWD blamed for fumes sickness." *The Daily News*, February 13, 1985: 1.

The Daily News. 1988. "Chemical waste are: 'Not a leaking dump, but a secure landfill', says IWD." *The Daily News*, March 17, 1988.

Tonkin & Taylor Limited. 2006. *Waireka Secure Containment Facility - Resouce Consent Variation: Assessment of Environmental Effects.* Prepared for Dow AgroSciences (NZ) Ltd. Wellington: Tonkin & Taylor.

United Nations Environmental Programme. 1996. Rotterdam Convention – Operation of the Prior Informed Consent Procedure for Banned or Severely Restricted Chemicals Decision Guidance Documents 2,4,5-T and Its Salts and Esters. Joint FAO/UNEP Programme for the Operation of Prior Informed Consent. http://www.pic.int/Portals/5/DGDs/DGD_2,4,5-T_EN.pdf

Wall, Tony. 2018. "It's banned in other countries but New Zealand is using more toxic methyl bromide than ever." *Stuff.co.nz*, June 4. Accessed August 30, 2020. https://www.stuff.co.nz/national/103690904/its-banned-in-other-countries-but-new-zealand-is-using-more-toxic-methyl-bromide-than-ever

Watts, Meriel. 1994. *The Poisoning of New Zealand.* Auckland, NZ: Auckland Institute of Technology Press.

Watts, Meriel. 2003. *Painted Apple Moth Eradication Programme: Health Risk and Effects.* https://www.semanticscholar.org/paper/PAINTED-APPLE-MOTH-ERADICATION-PROGRAMME-%3A-HEALTH-Watts/758c4f98b73b1e619 0d09facf8639763ee68a805

Weaver, Sean. 2003. "Policy Implications of 1080 Toxicology in New Zealand." *Journal of Rural and Remote Environmental Health* 2(2): 46–59.

Wildblood-Crawford, Bruce. 2008. "Environmental (in)justice and 'expert knowledge': The Discursive Construction of Dioxins, 2,4,5-T and Human health in New Zealand, 1940 to 2007." PhD Thesis, Department of Geography, University of Canterbury, 2008.

White, Alison. 2000. "Pesticides in food: why go organic?" *Pesticide Action Network NZ/Safe Food Campaign NZ. Safe Food Campaign Newsletter*, July/August. 2000.

5 Community Responses to the Spraying Operation

Prior to the aerial spraying operation there was significant community support for eradicating PAM, which included support from Bob Harvey, mayor of Waitakere City (which is a city that bordered Auckland and is where the moth was originally discovered); Green Party members of parliament; and some of the Community Advisory Group (CAG) that the Ministry of Agriculture and Forestry (MAF) had established (Walsh 2003). Additionally, a government survey of 600 residents revealed 86% supported eradication efforts (Office of the Ombudsman 2007). At the same time, however, many had serious qualms about using aerial pesticide spraying to achieve eradication, including some of the CAG and the handful of local residents who expressed concern at the outset and who were joined by others over the course of the operation.

Sources of Opposition: The Community Advisory Group

An important source of resistance came from the CAG, which MAF established in September 2001. The group included representation from interest groups (such as the Asthma society), doctor, teacher, three entomologists (including two who had played major roles in the 1996–1997 Project Ever Green), citizens who had lived through the 1996–1997 East Auckland spraying operation, and politicians, including an official from the Pesticide Board and the Agrichemical Trespass Ministerial Advisory Committee. The group's mandate consisted of being a conduit of information from and to MAF, which included conveying MAF's information to the community as well as conveying community concerns and suggestions to MAF (Walsh 2003; Watts & Blackmore 2002).

From its inception the group pushed for alternatives to aerial pesticide spraying, which included removing wattle (a tree species that painted apple moths were particularly attracted to), implementing vegetation removal controls to prevent residents from inadvertently transporting infested plant matter to other parts of the city, employing students to help locate infestations, and hiring a local entomologist (John Clearwater, who was a world-expert on pheromones and played a key role in ending the 1996–1997 white tussock moth infestation) to develop a pheromone that would attract male moths to traps (Walsh 2003).

DOI: 10.4324/9780429426414-6

The group also sought to publicly pressure government officials through the media. For instance, in November 2001 the group's chair (Kubi Witten-Hannah) publicly criticized MAF for having failed, 2.5 years into the eradication, to produce an operational plan and called on the Biosecurity Minister (Jim Sutton) to visit the West Auckland community (New Zealand Labour Party 2001). Additionally, after three months of trying to work with the spraying operation leadership, on December 17, 2001 (which was one month before the start of aerial spraying), the group publicly denounced that leadership for its failure to contain the moth and for allowing it to spread beyond its original infestation site (PAM CAG 2001). Moreover, they called for the PAM Operation Director (Ruth Frampton) to resign, as well as for disbanding and reconstituting the MAF Technical Advisory Group, which had been advising Frampton and had, thus, contributed to the failures (*Ibid.*). In another example, in June 2002, after six months of aerial spraying, the CAG called on the Prime Minister (i.e. Helen Clark) to immediately halt the airplane aerial spraying, in favor of a community-based approach that included intensifying ground spraying with Foray 48B and only using helicopters to target spray inaccessible areas (Green Party June 2002; PAM CAG 2002). Moreover, when Helen Clark was unresponsive to their suggestion, they publicly disclosed their efforts to persuade her by disseminating a press release to that effect (PAM CAG 2002).

MAF also had to contend with the individual efforts of two CAG members who produced knowledge that undermined MAF efforts. One was Hana Blackmore, whose February 2003 community health monitoring report catalogued the health complaints local residents were experiencing as a result of the spraying. Beyond revealing the litany of health issues that emerged over the course of the spraying, the report also revealed that hundreds of residents were experiencing health problems. The other individual was Meriel Watts, whose January 2003 report sought to explain the discrepancy between the Auckland District Health Board's 2002 Health Risk Assessment (HRA) and what was being experienced in the community. Towards that end, Watts analyzed the methodology that government officials used to produce the 2002 HRA and exposed its substantial flaws, which included a failure to properly account for exposure, ignoring the pesticide's neurological effects, the failure to consider how people would come in contact with the spray (i.e. dermal, inhalation, etc), and much more.

In turn, the release of these reports added to the growing pressure on the Ministry of Health to assess the spraying's impact on human health. In March 2003 the Ministry of Health finally succumbed to the pressure and commissioned researchers from the Wellington Medical School to investigate the issue. (Office of the Ombudsman 2007). However, while they commissioned the research, the Ministry also significantly undermined the process by circumscribing the project's scope, delaying the production of knowledge, and actively working to suppress its release. These are issues I explore more deeply in Chapter 8.

Sources of Opposition: Activist Groups

Besides the Community Advisory Group, opposition also came from numerous activist groups, including West Aucklanders Against Aerial Spraying (WASP), Society Targeting Overuse of Pesticides (STOP), the Sprayfree Coalition, Stop Aerial Spraying (SAS), Painted Apple Moth Community Coalition, Group Against Spraying, Mothers Against Spraying Kids, Spray Action Group, and the Pesticide Action Network New Zealand (PAN NZ).

Like the Community Advisory Group, the activist groups publicly criticized numerous aspects of MAF's spraying operation. For instance, STOP (2002) criticized MAF leadership for failing to consider community concerns about the eradication, its lack of transparency, and, ultimately, its failure to contain and eradicate the moth. Given its grievances, a month before the start of aerial spraying (i.e. December 17, 2001), the group also supported the Community Advisory Group's call for Ruth Frampton's resignation. (*Ibid.*).

A second prominent activist group was WASP, which criticized MAF for (1) failing to properly communicate upcoming spraying events to residents; (2) failing to deploy sterile moths earlier in the eradication operation; and (3) requiring home owners and tax payers to pay for replacing the trees and other plant life that MAF destroyed as part of its eradication operation (2002a, 2002c, 2002d). In press releases the group's leader (Helen Wiseman-Dare) also cast doubt on MAF's claim that PAM was capable of destroying trees by sharing the following at the November 30th rally:

> The woman who first discovered the moth on her property and called in the local entomologist who took it to MAF, said she is aghast by what has happened. She said the affected trees on her property all regenerated. We've seen the same with the infested Wattles on Traherne Island at the beginning of the year before spraying started.
>
> (WASP 2002f)

WASP also criticized the Public Health Service for its unwillingness to carry out proper longitudinal studies to assess the long-term effects of the spray (WASP 2002b)

Besides publicly criticizing government agencies, the groups conducted a range of other activities. For instance, WASP disseminated information through leafletting, writing letters to the editor, giving media interviews, disseminating press releases, and lobbied members of Parliament and local politicians through letter-writing and in-person visits. As well, the group gathered signatures for petitions, organized a letter-writing campaign to the Prime Minister (Helen Clark), and, to name and shame those advising MAF's operation, publicly revealed the names and e-mail addresses of those on the PAM operation's Science and Technical Advisory Group (WASP 2002f).

Additionally, the activist groups organized numerous protest marches, rallies, and picketing, including those occurring on October 12, 2002, October 23, 2002, November 9, 2002, November 30, 2002, and January 20, 2003, January 28, 2003, March 29, 2003, and July 8, 2003, (No Way Spray 2003;

Painted Apple Moth Community Coalition 2003; Sprayfree Coalition 2002; Stop Aerial Spraying 2002; Tyson 2009; WASP 2002e, 2003a, 2003b). In addition, the Stop Aerial Spraying group sought a legal opinion from Sir Geoffrey Palmer (lawyer and former New Zealand Prime Minister) regarding the potential for seeking an injunction under the country's Health Act (Green Party 2003b).

Politicians: Latecomers to the Opposition

Opposition also emerged from some elected officials, though their opposition arrived quite late in the process. Moreover, their opposition was uneven, particularly at the beginning, as all of them unequivocally supported eradication efforts at the operation's start and only began opposing the spraying when MAF missed their initial targets or substantially increased the operation.

At the local level, while the Waitakere City Council manifested an initial support for eradication efforts, after the operation's substantial expansion (in October 2002) local politicians provided increasing support for opposition efforts. For example, in November 2002 the 18-member Waitakere City Council continued providing operational support (access to meeting rooms, copiers, etc) to the CAG that MAF had recently disbanded for being too critical of the spraying operation (Green Party Nov 7 2002; PAM CAG 2002b). Additionally, in December 2002 the Waitakere City mayor (i.e. Bob Harvey) lobbied MAF to (1) provide full medical support to those who were falling ill from the spraying; (2) reinstate the CAG that had recently been dissolved; and (3) publicly release the list of ingredients in the Foray 48B pesticide (Walsh 2003). As well, that same month the City Council approved a $10,000 payment to the Stop Aerial Spraying activist group, in order to fund their efforts to have Geoffrey Palmer (prominent human rights lawyer and former New Zealand Prime Minister) investigate whether the health effects of the aerial spraying warranted pursuing legal recourse via the country's Health Act (*Ibid.*).

Having said that, it needs to be underscored that the local politicians did not start supporting opposition efforts until ten months into the spraying operation. Moreover, after Geoffrey Palmer declared that the government's use of the Biosecurity Act did not supersede the country's Health Act and that local authorities who were unsatisfied with the safety of the aerial spraying "were empowered to take all steps to secure the abatement of the nuisance," the Waitakere City Council failed to pursue the matter any further (Thompson 2003).

Green Party members of Parliament were another source of opposition, though its opposition was also uneven and arrived very late in the process. Like the local officials, Green Party officials unequivocally supported eradication efforts at the outset. Beyond simply supporting eradication efforts, they unwittingly supported government manipulation efforts by parroting MAF's unsupported narrative that PAM would be a menace to society.

For example, in different press releases the party's biosecurity spokesperson (MP Ian Ewen-Street) referred to the moth as a "voracious feeder". He stated that its spread "could be disastrous for native biodiversity," and that having to live with the moth could "devastate both native bush and logging plantations" (Green Party 2001c, 2001d, 2002a).

Although the Green Party did criticize MAF at the outset, these criticisms were always centered on MAF's failure to effectively eradicate the moth, with their Biosecurity spokesperson repeatedly criticizing the agency for its ineptness at containing and eradicating PAM (Green Party 2001a, 2001b, 2001c, 2001d, 2002b, 2002j, 2004a, 2004b). Additionally, he repeatedly called for an official government inquiry to investigate how MAF had allowed PAM to spread so extensively (Green Party 2001d, 2001e, 2002c). Along these lines, in March 2002 he succeeded in forcing MAF to meet with a parliamentary committee (i.e. the Primary Production Select Committee) in order to publicly explain why, after nearly three years, the agency had yet to eliminate the moth (Green Party 2002b). Furthermore, in April 2004, as the spraying operation was winding down, he pressed for a parliamentary inquiry to investigate MAF's handling of the whole operation (Green Party 2004b).

Although party officials were emphatically critical of MAF's ineffectiveness, their communications around the aerial spraying were quite problematic. For instance, while the Green Party co-leader (Jeanette Fitzsimons) criticized MAF for its initial ground spraying with chlorpyrifos (which is known to negatively impact human and environmental health (ATSDR 1998; CHE 2020; Rauh et al. 2011), she actually advocated replacing that practice with aerial spraying of Btk (Green Party 1999). To be fair, in an October 2001 press release (which came out three months before the start of aerial spraying), the Biosecurity spokesperson did give voice to citizen concerns about the spraying through the following statement:

> Spraying any substance over suburban housing should be a very last resort. I am concerned at the possible long-term health effects and there is anecdotal evidence that there were health effects from the last Btk spraying for the White Spotted Tussock moth in Eastern Auckland.
> (Green Party 2001a)

However, this "concern" was undermined in the next paragraph where he stated "We reluctantly accept that Btk aerial spraying is now the only remaining option," which tacitly endorsed MAF's decision to proceed with aerial spraying. This tacit support was reiterated in press releases disseminated the next two months, which argued there is "no option for elimination other than aerial spraying" (Green Party 2001c) and "There now seems to be no alternative to spraying on a wide scale" (Green Party 2001e).

In another example of weak opposition, a full month into the spraying the Green Party Biosecurity spokesperson criticized MAF for refusing to divulge the ingredients in the pesticide: "The ingredients of the spray used, Foray

48B, have been kept a secret for commercial reasons. But surely people have a right to know what they and their children are being sprayed with?" (Green Party 2002a). However, this issue then lay dormant for seven months, until other Green Party officials revisited the issue in September and October 2002, as the spray zone was being significantly expanded (Green Party 2002h, 2002i)

June 2002 was a turning point for the Green Party as that is when party officials finally provided unequivocal opposition to the aerial spraying, with MP Ewen-Smith stating:

> The whole Painted Apple Moth incursion has been terribly misman-aged by MAF since day one. Residents should not have to be repeatedly sprayed because of MAF's bungling of this incursion and it is now time to address the health concerns that many residents hold

and "The Greens believe it is time for a new approach that does not subject residents to more blanket spraying" (Green Party 2002d). For context, this about face came a month after MAF revealed it had failed to eradicate the moth by the original end date of May 2002 and would continue spraying indefinitely. From that point onwards, Green Party communications (2002e, 2002f, 2002g, 2003d) consistently emphasized community health concerns. Additionally, in February and March 2003, after the release of the Blackmore and Watts reports, the party's health spokesperson (Sue Kedgley) renewed attention to the government's unwillingness to divulge the spray ingredients, while also calling for an urgent independent investigation of the health effects associated with the spray (Green Party 2003b, 2003c). Moreover, in May of that year she publicized that some activists had identified some of the spray ingredients (Green Party 2003e).

Besides raising health concerns, Green Party officials repeatedly provided support to those calling to move away from aerial spraying. For instance, in a June 2002 press release Ewen-Street supported the Community Advisory Group's plea to abandon aerial spraying in favor of intensified ground spray-ing with Foray 48B (Green Party 2002d). Moreover, a week later he chastised MAF when it decided to ignore CAG's suggestion and to press on with aerial spraying (Green Party 2002e). As well, in January 2003 he lent support to the Waitakere City major (i.e. Bob Harvey), when he too pressed MAF to cease aerial spraying in favor of ground spraying (Green Party 2003a).

Why Was MAF Able to Carry Out a Spraying Operation of Unprecedented Size and Duration?

Given that MAF's spraying operation faced some opposition at the start of the operation and that the opposition grew over time, particularly after MAF de-cided to extend the operation in May 2002, one needs to ask why they were still able to carry out an aerial pesticide spraying operation of unprecedented scope and duration. This section reveals two major factors that contributed to that outcome, with one being the use of a legal loophole and the other being that while MAF faced initial opposition at the beginning, that opposition was not widespread and it grew slowly.

Using a Legal Loophole to Sidestep Local Opposition

The country's Resource Management Act (RMA) normally provides a mechanism through which local populations can object to pesticide spraying activities pursued by the central government. Specifically, the RMA states that "it is an offence to discharge any contaminants into the environment unless the discharge is expressly allowed by a rule in the regional plan and any relevant proposed regional plan, a resource consent, or by regulations" (Office of the Ombudsman 2007: 103). This gives local citizens an important mechanism to exercise oversight over spraying activities being pushed by the central government.

However, in 1997, following the 1996–1997 spraying operation in East Auckland, government officials created a loophole for themselves, by adding a new section to the Biosecurity Act that allowed the Minister of Agriculture and Forestry to exempt biosecurity actions from the RMA regulations when they meet the following threshold criteria:

> (A) The organism is not established in New Zealand, the organism is not known to be established in New Zealand, or the organism is established in New Zealand but is restricted to certain parts of New Zealand; and (B) The organism has the potential to cause all or any of significant economic loss, significant adverse effects on human health, or significant environmental loss if it becomes established in New Zealand or if it becomes established throughout New Zealand; and (C) It is in the public interest that action be taken immediately in an attempt to eradicate the organism.
> (Office of the Ombudsman 2007: 103)

During the painted apple moth operation MAF used this loophole to effectively override local ordinances that could have otherwise prevented the spraying (*Ibid.*).

The Limited Spread of Opposition amongst the Masses

Although the Biosecurity Act loophole was certainly an important legal mechanism for overriding local legal challenges, another factor that needs to be considered is the spread of public opposition to the spraying. Such opposition matters because pesticide use is associated with numerous health and environmental concerns, which can lead to mass mobilization for increasing regulations or even banning pesticide use. An example is the province of Québec (Canada), where mobilizations led to the banning of multiple pesticide products (Hernke & Podein 2011; Rickman 2004).

As this pertains to the PAM operation, the case was marked by an initial limited opposition, which spread slowly. While there was certainly a core of people who presented unequivocal opposition to the spraying at the beginning of the operation, which grew to eventually become a catalyst for a public inquiry and an investigation by the Office of the Ombudsman, the initial opposition was quite limited. This is indicated by a government-commissioned survey showing 62% of Aucklanders agreed with using targeted aerial spraying to eradicate the moth with a significant portion of those "strongly" agreeing (19% agreed and 43% strongly agreed), and only 11% disagreed (MAF 2001). Moreover, the opposition grew quite slowly, as underscored by the fact that support for the spraying was still quite high ten months into the operation (i.e. November 2002), with another survey showing that 67% of participants agreed with the statement "the eradication programme is inconvenient, but ultimately worth it to stop the environmental damage and health effects the moth could cause," with only 13% of residents disagreeing with the statement (Office of the Ombudsman 2007: 50).[1]

The limited opposition in November 2002 is even more startling when we consider that, by that point, MAF had already carried out pesticide spraying operations on 20 different days (Goven et al. 2007). This amount of spraying was already far more than the six to eight sprayings MAF had originally proposed to administer over the area. Additionally, when the survey was taken MAF had already considerably expanded the spray zone from the 300 hectares (741 acres) that were originally proposed to 7,980 hectares (19,719 acres) (Goven et al. 2007). Moreover, by that point MAF had publicly announced their intention to carry out more sprayings in the weeks and months to come. Interestingly, the November survey also revealed that even though support was weaker among residents in the "hot spot" areas (which were areas that experienced the most spraying), even in those areas 51% of residents believed MAF was doing a good job of eradicating the moth, with only 27% disagreeing (Office of the Ombudsman 2007: 49–50).

The slow spread of opposition was also indicated by the low attendance at protests. For example, a January 2002 protest outside MAF's base of operations (in the suburb of Henderson) only yielded 50 protestors, despite the fact an estimated 13,500 people were living in the spray zone (Blackmore 2020; TVNZ 2007). Similarly, the following October only 80 people showed up for a protest outside the Prime Minister's Auckland home, despite the fact the spray area had been dramatically expanded from 962 hectares (2,377 acres) in September to 7,980 hectares (19,360 acres) (Goven et al. 2007; TVNZ 2007; Tyson 2009). While 600 and then 1,500 participated in October and November protest marches, that was just a fraction of the over 150,000 people who had been exposed to the spraying or the 200,000+ who would be exposed to the spraying within the subsequent five weeks (*Ibid.*).

The opposition's slow spread was significant for MAF as it meant it was unlikely to face significant political repercussions for prolonging the operation and expanding its scope. While opposition continued to build up over time, it didn't reach a critical mass prior to MAF completing its 29-month spraying operation.

Potential Concerns about the Spraying Operation

The slow spread of opposition is surprising when we consider the substantial concerns that existed around this spraying operation. First, when MAF first announced its intention to administer aerial pesticide spraying it was clear the pesticides would be sprayed in residential neighborhoods, not a thinly populated agricultural field that would have been out-of-sight and out-of-mind. It is one thing for urbanites to accept the use of pesticides to grow food in the hinterlands, but quite another to accept a spraying operation in their own backyard, where they will viscerally experience the spraying, by seeing the pesticides being sprayed; smelling the pesticide in their gardens, homes, and clothing; feeling the pesticide's burn on their skin and in their lungs; and hearing the helicopters and planes circling back for another pass.

Another factor that could have concerned West Aucklanders was the pesticide spraying campaign that was conducted in East Auckland in 1996–1997, to eradicate the white-spotted tussock moth (Office of the Ombudsman 2007). That operation went on for eight months and exposed over 86,000 people to the spraying. One imagines that, only four years later, West Aucklanders might have had a vivid recollection of what the previous operation entailed and would have opposed allowing something similar to unfold in their backyards.

Third, Auckland residents could have been concerned about the human health effects reported after previous Foray 48B sprayings, both in East Auckland and in North America. For instance, over 300 East Aucklanders traced health problems to the 1996–1997 spraying (Blackmore 2003). One imagines those casualties would have given West Aucklanders even more concerns about allowing a similar operation to be carried out in their neighborhoods.

Fourth, West Aucklanders were kept in the dark about all the ingredients that were in the pesticide. It is one thing to accept being exposed to a substance when you know exactly what that substance is and what potential problems it can cause. However, it is quite another matter to be exposed to a pesticide whose ingredients are withheld from you, which prevents you from ensuring your and your family's safety, and can be quite unsettling. While the WASP community group protested the ingredient secrecy from the outset, the masses seemed unperturbed by the lack of disclosure.

Fifth, West Aucklanders could have been disturbed by the number of times the pesticide was administered. Although the government initially intended to limit the operation to six to eight sprayings, by the end of the operation they carried out aerial sprayings on at least 60 days, with double passes (one by airplane and another by helicopter) on many of those days. It is one thing for a population to accept a pesticide spraying operation of limited duration (i.e. 6 to 8 sprayings over 4 months) but quite another to continue tolerating a spraying operation that persists month after month, over the span of nearly two and a half years. Again, while a small group of activists protested the sprayings right from the outset, this was not emulated by the masses.

Summary

This chapter focused on how the community responded to the spraying operation, which included identifying sources of opposition as well as discussing the slow spread of that opposition to the masses. Although it is important to note the slow spread of opposition, it behooves us to understand why opposition to the spraying spread so slowly, particularly when we consider the potential concerns citizens could have had about the spraying.

Towards that end, it's important to understand that levels of public support and opposition do not occur in a vacuum but rather are shaped by the efforts of social forces, such as chemical companies, government agencies, and activist groups (Cockburn 2011). Government agencies played a central role in shaping support for the PAM spraying operation, and the next chapter focuses on how they sought to build public support by using a fear-based communication campaign that framed PAM as a triple biosecurity threat: a threat to the nation's economy, native ecology, and public health.

Note

1 It needs to be specified that the vast majority of the people surveyed were outside the spray zone, and so were unlikely to know the degree to which people in the spray zone were being impacted.

References

Agency for Toxic Substances and Disease Registry (ATSDR). 1998. "Public Health Statement Chlorpyrifos." Toxic Substances Portal, US Government. https://www.atsdr.cdc.gov/ToxProfiles/tp84-c1-b.pdf

Auckland District Health Board, Public Health Service. 2002. *Health Risk Assessment of the 2002 Aerial Spray Eradication Programme for the Painted Apple Moth in some Western Suburbs of Auckland.*

Blackmore, Hana. 2003. *Interim Report of the Community-based Health and Incident Monitoring of the Aerial Spray Programme: January–December 2002.* Auckland February 2003.

Cockburn, Andrew. 2015. "Weed Whackers: Monsanto, Glyphosate, and the War on Invasive Species." *Harper's Magazine,* September: 57–63.

Collaboration on Health and the Environment (CHE). 2020. *Toxicants and Disease Database: Chlorpyrifos.* https://www.healthandenvironment.org/our-work/toxicant-and-disease-database/?showcategory=&showdisease=&showcontaminant=2696&showcas=&showkeyword=

Goven, Joanna, Tom Kerns, Romeo Quijano, and Dell Wihongi. 2007. *Report of the March 2006 People's Inquiry Into the Impacts and Effects of Aerial Spraying Pesticide over Urban Areas of Auckland.* Auckland, NZ: Action Plan & Print. https://peoplesinquiry.wordpress.com/

Green Party. 1999. "MAF takes toxic option to control moth." *Scoop Independent News,* October 10. https://www.scoop.co.nz/stories/PA9910/S00170.htm

Green Party. 2001a. "Moth spraying could well have been avoided." *Scoop Independent News,* October 29. https://www.scoop.co.nz/stories/PA0110/S00488.htm

Green Party. 2001b. "MAF blow chance with painted apple moth – Greens." *Scoop Independent News,* November 9. https://www.scoop.co.nz/stories/PA0111/S00165.htm

Green Party. 2001c. "MAF must be accountable for spread of moth." *Scoop Independent News,* November 28.

Green Party. 2001d. "Greens want inquiry into MAF handling of Moth." *Scoop Independent News,* December 04. https://www.scoop.co.nz/stories/PA0112/S00047.htm

Green Party. 2001e. "MAF Have No Idea on Painted Apple Moth." *Scoop Independent News,* December 14. http://www.scoop.co.nz/stories/PA0112/S00266.htm

Green Party. 2002a. "Greens to grill MAF over moth fiasco." *Scoop Independent News,* February 26. https://www.scoop.co.nz/stories/PA0202/S00385.htm

Green Party. 2002b. "Greens seriously concerned with MAF's explanations." *Scoop Independent News,* March 15. https://www.scoop.co.nz/stories/PA0203/S00288.htm

Green Party. 2002c. "Inquiry Into Painted Apple Moth Cannot Be Avoided." *Scoop Independent News,* May 7. http://www.scoop.co.nz/stories/PA0205/S00150.htm

Green Party. 2002d. "Change of approach needed on Painted Apple Moth." *Scoop Independent News,* June 25. https://www.scoop.co.nz/stories/PA0206/S00431.htm

Green Party. 2002e. "Government Kicking for touch on Painted Apple Moth." *Scoop Independent News,* July 3. https://www.scoop.co.nz/stories/PA0207/S00119.htm

Green Party. 2002f. "Painted Apple Moth." *Scoop Independent News,* July 4. http://www.scoop.co.nz/stories/PO0207/S00052.htm

Green Party. 2002g. "Greens Challenge Govt to List Spray Ingredients." *Scoop Independent News,* September 11.

Green Party. 2002h. "Greens challenge Govt to list spray ingredients." *Scoop Independent News,* September 13. https://www.scoop.co.nz/stories/PA0209/S00227.htm

Green Party. 2002i. "Greens challenge Govt to list spray ingredients." *Scoop Independent News,* October 7. https://www.scoop.co.nz/stories/PA0210/S00134.htm

Green Party. 2002j. "Painted Apple Moth Vegetation Restrictions." *Scoop Independent News,* December 10. https://www.scoop.co.nz/stories/PA0212/S00225.htm

Green Party. 2003a. "Back to basics on the painted apple moth." *Scoop Independent News,* January 10. https://www.scoop.co.nz/stories/PA0301/S00048.htm

Green Party. 2003b. "Moth spray needs health check-up." *Scoop Independent News,* February 4. https://www.scoop.co.nz/stories/PA0302/S00032.htm

Green Party. 2003c. "Full health study needed on moth spray." *Scoop Independent News,* March 14. https://www.scoop.co.nz/stories/PA0303/S00242.htm

Green Party. 2003d. "Green MPs put spotlight on PAM spray campaign." *Scoop Independent News,* April 3. https://www.scoop.co.nz/stories/PA0304/S00068.htm

Green Party. 2003e. "DIY detectives reveal 'secret' spray ingredients." *Scoop Independent News,* May 5. https://www.scoop.co.nz/stories/PA0305/S00087.htm

Green Party. 2004a. "Moth paints picture of perpetual MAF failure." *Scoop Independent News*, January 21. https://www.scoop.co.nz/stories/PA0401/S00147.htm

Green Party. 2004b. "Inquiry needed into MAF's moth inaction." *Scoop Independent News*, April 20. https://www.scoop.co.nz/stories/PA0404/S00300.htm

Hernke, Michael T., and Rian J. Podein. 2011. "Sustainability, Health and Precautionary Perspectives on Lawn Pesticides, and Alternatives." *EcoHealth* 8: 223–232.

Ministry of Agriculture and Forestry (MAF). 2001. "Majority Support Targeted Aerial Spraying." *Scoop Independent News*, December 18. https://www.scoop.co.nz/stories/GE0112/S00035.htm

New Zealand Labour Party. 2001. "Moth Problem Needs Ministerial Visit – Cunliffe." *Scoop Independent News*, November 26. https://www.scoop.co.nz/stories/PA0111/S00429.htm

No Way Spray. 2003. "Spray Group Pressure Local MP." *Scoop Independent News*, January 20. https://www.scoop.co.nz/stories/PO0301/S00053.htm

Office of the Ombudsman. 2007. *Report of the Opinion of Ombudsman Mel Smith on Complaints Arising from Aerial Spraying of the Biological Insecticide Foray 48B on the Population of Parts of Auckland and Hamilton to Destroy Incursions of Painted Apple Moths, and Asian Gypsy Moths, Respectively.* Wellington, NZ: Office of the Ombudsman.

Painted Apple Moth Community Advisory Group (PAM CAG). 2001. "Get Ruth-Less with the Painted Apple Moth." *Scoop Independent News*, December 17. https://www.scoop.co.nz/stories/PO0112/S00077.htm

Painted Apple Moth Community Advisory Group (PAM CAG). 2002a. "Halt to aerial spray programme called for." *Scoop Independent News*, June 24. https://www.scoop.co.nz/stories/PO0206/S00151.htm

Painted Apple Moth Community Advisory Group (PAM CAG). 2002b. "Community Advisory Group Harder to Get Rid of." *Scoop Independent News*, November 8.

Rauh, Virginia, Srikesh Arunajadai, Megan Horton, Frederica Perera, Lori Hoepner, Dana Barr, and Robin Whyatt. 2011. "Seven-Year Neurodevelopmental Scores and Prenatal Exposure to Chlorpyrifos, a Common Agricultural Pesticide." *Environmental Health Perspectives* 119(8): 1196–1201.

Rickman, Angela. 2004. "Canadian Activists Win Pesticide Bylaws." *Global Pesticide Campaigner* 14(2): 1.

Society Targeting Overuse of Pesticides (STOP). 2002. "S.T.O.P. Supports Call for MAF Resignations." *Scoop Independent News*, December 17.

Sprayfree Coalition. 2002. "Aerial Spray Protest March Sat. 30th November." *Scoop Independent News*, November 28. https://www.scoop.co.nz/stories/AK0211/S00134.htm

Stop Aerial Spraying. 2002. "Government Refused to Receive Community Concerns." *Scoop Independent News*, November 15. https://www.scoop.co.nz/stories/PO0211/S00120.htm

Thompson, Wayne. 2003. "Moth spray may be illegal, says Palmer." *New Zealand Herald*, January 31, 2003.

TV New Zealand (TVNZ). 2007. "Painted apple moth timeline." Accessed October 19, 2015. http://tvnz.co.nz/content/1443394/2591764/article.html

Tyson, Janet. 2009. "The Painted Apple Moth Eradication Programme (vignette version: A)." *The Case Program, The Australian and New Zealand School of Government.* www.anzsog.edu.au

Walsh, Frances. 2003. "Wipe Out." *Metro* 261: 44–45.

Watts, Meriel and Hana Blackmore. 2002. *Report and Recommendation of Special Science Meeting: Painted Apple Moth Eradication – Community Option – (MAF Option 3)* May 30.

West Aucklanders against Aerial Spraying (WASP). 2002a. "BTK Spraying in Middle of Busy Shopping Centre." *Scoop Independent News*, June 17. https://www.scoop.co.nz/stories/AK0206/S00051.htm

West Aucklanders against Aerial Spraying (WASP). 2002b. "Painted Apple Moth Aerial Spraying." *Scoop Independent News*, July 4. https://www.scoop.co.nz/stories/PO0207/S00053.htm

West Aucklanders against Aerial Spraying (WASP). 2002c. "Ratepayers to foot the bill for Painted Apple Moth." *Scoop Independent News*, September 6. https://www.scoop.co.nz/stories/AK0209/S00008.htm

West Aucklanders against Aerial Spraying (WASP). 2002d. "Govt Puts Health of Forest Before West Aucklanders." *Scoop Independent News*, September 10. https://www.scoop.co.nz/stories/PO0209/S00037.htm

West Aucklanders against Aerial Spraying (WASP). 2002e. "West Aucklanders Reel from BTK Laden Planes." *Scoop Independent News*, October 30. https://www.scoop.co.nz/stories/AK0210/S00138.htm

West Aucklanders against Aerial Spraying (WASP). 2002f. "Helen Wiseman–Dare Speech: Anti Spray March Nov. 30." *Scoop Independent News*, December 4. https://www.scoop.co.nz/stories/AK0212/S00023.htm

West Aucklanders against Aerial Spraying (WASP). 2003a. "Safe Non-toxic Pest Control Must be Fast Tracked." *Scoop Independent News*, March 27. https://www.scoop.co.nz/stories/AK0303/S00138.htm

West Aucklanders against Aerial Spraying (WASP). 2003b. "Anti Aerial Spray Protest 9:30 am July 8." *Scoop Independent News*, July 7. https://www.scoop.co.nz/stories/AK0307/S00045.htm

6 Framing Foreign Species as Biosecurity Threats

Although the use of any pesticide should raise concerns about human and ecosystem health, most urban residents are disengaged from such concerns. To some degree, this disengagement stems from the fact most pesticide spraying is administered for agricultural purposes and takes place in thinly populated rural areas, where it is out-of-sight and out-of-mind to city dwellers. However, one imagines it would be a different story with aerial pesticide spraying conducted in urban areas, where residents will viscerally experience the spraying, either through seeing and hearing the approaching planes and/ or helicopters, seeing the pesticide being sprayed, feeling it on their skin, smelling its odor, or experiencing illness symptoms afterwards.

One also imagines this visceral experience would prompt citizens to be concerned about potential health risks, which could, in turn, generate significant opposition and potential political backlash. This is particularly true in Western democracies, where there is a strong recent record of movements building around ecological issues. However, in the case of the PAM spraying, the opposition was slow to spread. What makes the delay even more notable is that there were numerous potential concerns associated with the spraying, not the least of which were the health problems that had emerged in previous spraying operations.

In trying to account for the slow spread of opposition, this chapter draws on Laura Nader's (1997) "controlling processes" concept, which she defines as micro-processes through which certain conditions become normalized and through which "individuals and groups are influenced and persuaded to participate in their own domination" (p. 712). She argues that such processes are particularly prevalent in industrialized democratic societies, where coercive power is less culturally acceptable and where power is increasingly exerted through cultural controls, which channel taste, values, and behavior.

Nader's work encourages us to identify the social forces operating in a given context and to analyze the way their actions curtail potential resistance. As it pertains to urban pesticide spraying operations, government agencies are a significant social force, as they are either the main proponent of such spraying activities or enable industrial agents to pursue such spraying. Given the government's central role, it behooves social scientists to identify how

DOI: 10.4324/9780429426414-7

government actions, whether intentional or not, help reduce the spread of opposition to pesticide use.

One way pesticide proponents have been known to slow opposition is to build public support for pesticide spraying. Moreover, as Andrew Cockburn (2015) elucidates, one way they have done so is by heightening the perceived risks associated with foreign species. This is exemplified well in the PAM case, where government officials deployed a fear-based, million dollar communication campaign that portrayed PAM as a significant biosecurity threat to the country (Beston 2001; PAM CAG 2002). In order to stimulate citizen fears, Ministry for Agriculture and Forestry (MAF) officials sought to emphasize the damage the moth *could* do, so that people would come away with the belief that while the spraying might be inconvenient, it was necessary to protect the country (Tyson 2006).

In disseminating this message MAF officials used various communication modalities, including direct-mail pamphleting, sending out press releases that got picked up by mainstream media, and purchasing advertising on billboards, radio, television, and newspapers, which included full-page adverts in the *New Zealand Herald*, Auckland and New Zealand's largest newspaper (Tyson 2006). Consequently, residents would read about the moth in the morning newspaper, after which, on their way to work, they would see the moth much larger than life on highway billboards and hear about it on the car radio (MAF purchased advertising on seven different radio stations) while they idled in Auckland's legendary traffic (*Ibid.*). They would read about it again when they rifled through their mail at the end of the day, perhaps catch the message on a fridge magnet that MAF sent out to residents, and then catch hard-hitting TV advertising in the evening (*Ibid.*). As one informant remarked, "the messaging was everywhere and one couldn't get away from it." Moreover, the hard-hitting TV ads were particularly effective as "experienced marketers were astounded to find the PAM enjoyed almost 100% brand recognition" (Tyson 2006: 9).

To better analyze the state's communication work I draw on Gary Alan Fine's (1988) "naturework" concept, which refers to the ideological work associated with converting nature into culture. Fine (1988) starts with the premise that our perceptions of things in nature (such as trees, animals, and rivers) are not inherent to those things, but rather are socially constructed, through the daily ideological work of ascribing social meanings to those things, which effectively transforms "nature" into culture. Moreover, he emphasizes this naturework is heavily inflected by our cultural context, which conditions us to apply certain social meanings as opposed to others. For instance, whether we primarily view trees as a source of oxygen, a source of medicine, an integral part of ecosystems, or simply a source of building materials will be significantly mediated by the meanings and values our culture attributes to trees. The concept can also be extended to foreign species, for how we view them will also be significantly mediated by our cultural conditioning. If we live in a society that conditions us to view foreign species as threats, we are much more likely to see a foreign species as a potential problem than would be the case for someone living in a society where foreign species are less problematized.

Importantly, while our perceptions of nature are socially constructed, the power to shape those perceptions is not equally shared by all. This point is illustrated in Stella Capek's (2009) discussion of the naturework carried out by car manufacturers, who powerfully frame nature as something waiting to be conquered by humans in motorized vehicles. Besides corporations, government agents also enjoy substantial power to shape public perceptions of nature, due to the resources they possess to develop frames that support their political agendas and to the substantial access they have to the mass media system. As a case in point, the New Zealand government had the resources to allocate a million dollars towards their painted apple moth communications campaign, which relied substantially on mass media (Beston 2001).

This chapter examines ideological work that government officials pursued to have PAM viewed as a significant threat to society. In pursuing this analysis I systematically reviewed all PAM-related press releases that government agents disseminated via government websites, including the government's main websites (i.e. Beehive.govt.nz as well as those belonging to the Minister for Primary Industries and Biosecurity) and/or appeared on the *Scoop Independent News* website. I also considered statements that government officials made about PAM in other venues, including in parliamentary debates and in newspaper articles and advertisements. My analysis revealed that government communication activities encouraged the public to view PAM as a three-fold biosecurity threat: a threat to the country's economy, New Zealand's indigenous flora, and human health.

Framing PAM as an Economic Threat

One aspect of the government's framing was to portray PAM as a threat to the nation's economic interests. In 2000 MAF analysts predicted that, in the absence of government intervention, the PAM incursion would entail negative impacts on private amenity, public amenity, plantation forestry, horticulture, the conservation estate, watershed conservation, human health, and trade prospects. In particular, basing themselves on estimates produced for the 1996–1997 white-spotted tussock moth incursion, the analysts predicted the PAM incursion would cost the country $16–$116 million over the subsequent 20 years, with a likely total cost of $48 million (MAF 2000). In January 2002 MAF began the aerial spraying portion of the eradication operation and in May of that year, when it became clear the initial eight sprayings did not eradicate the moth and that more sprayings would be needed, MAF increased the impact estimates to $58–$356 million (MAF 2002b).

In turn, government officials used the "economic threat" framing at key moments of the spraying operation. For instance, in December 2001, a month before the start of spraying, MAF released a press release claiming PAM's spread would cost the country "at least $48 million over 20 years." The exact same claim was made in another press release the following month, when MAF began the operation's aerial spraying component (MAF 2002a). In September 2002, as MAF was about to significantly expand the spray zone (from 962 to 7,980 hectares), additional government press releases emphasized that the spraying was important to protect the country's forest industry (MAF 2002d, 2002e).

The frame was reiterated later in the operation, when the government came under increasing criticism. For instance, in March 2003 Judith Tizard (Auckland Central MP) declared, "the painted apple moth represents a projected cost to the economy of up to $358 million over the next 20 years" (Tizard 2003: 1). It is worth mentioning Tizard's statement came after the February release of Hana Blackmore's (2003) report, which documented community health problems associated with the spraying and which became a rallying cry for opposition efforts. Furthermore, in May 2004, at a time when the government was coming under political pressure for suppressing the Wellington Medical School's report on pesticide health effects, Jim Sutton (Biosecurity Minister) reiterated the economic frame "this moth is a serious threat to our urban, native and commercial trees with an estimated economic impact of about $258 million" (Sutton 2004: 1).

Framing a foreign species as a multi-million dollar threat is a potent way to stoke concern, if not fear, about that species' presence, which would have been essential for galvanizing support to have it eradicated, pursuing a pesticide spraying operation over urban areas, and getting the required operational funding.

Problems with the "Economic Threat" Frame

Although the "economic threat" frame was a potent means of generating political support, there were several problems with it.

First, while MAF analysts presented PAM as an economic threat to all of New Zealand, in reality it was the forestry sector that was most likely to be impacted. Specifically, in a low-impact scenario (i.e., $58 million), MAF estimated the plantation forestry sector would shoulder 78% of economic impacts ($45.2 million), which would increase to 87% in a high-impact scenario (i.e. $311.4 million out of $358 million) (MAF 2002b).

Second, the economic estimates were based on the questionable assumption that the moth would affect a far greater percentage of pine trees than in its natural habitat. When Australian radiata pine forests experience a PAM infestation, less than 1% of trees in an infested compartment are affected (MAF 2000). By contrast, MAF's analysts assumed the number of infested New Zealand trees would be 2–25 times the Australian rate. The use of such high estimates was not based on empirical evidence, but rather on the belief that New Zealand's lack of natural predators would enable the PAM population to infest a larger number of pine trees than occurs in Australian pine tree plantations. In turn, that reasoning was based on the assumptions that (A) the moth has natural predators in Australia; (B) it is these predators that restrict the moth's spread in that country; and (C) New Zealand ecosystems would have no resistance to the moth. Although a case could plausibly made for the first two assumptions, MAF failed to make the case for any of those assumptions. Additionally, not only did MAF fail to make the case for the third assumption, that assumption was at odds with the Island Resource Allocation Hypothesis, which suggests such invaders have a hard time settling in isolated island floras, such as New Zealand (see the next section for more information about this hypothesis) (Suckling et al. 2014). As well, there is evidence to suggest painted apple moths were infected by local organisms, including fungi and parasitic wasps (WASP 2002).

Another questionable assumption was the analysts' belief that trees infested with the moths would experience significant defoliation, which would stunt their growth (MAF 2002b). In particular, analysts estimated trees affected by PAM would suffer a 60% reduction in growth (*Ibid.*). This assumption was not based on empirical data related to PAM infestations in Australia. Rather, the estimate was based on losses experienced following an infestation of Cyclaneusma needle cast, which is a pine tree fungal infection (*Ibid.*). No justification was given for using the fungal infection as the basis for estimating the defoliation that PAM might cause. Another problem with the analysts' assessment is the fact pine trees were not PAM's preferred host, as the moths were predominately found on wattles, which is a type of weed (Forest Research 1999; Frampton 2002). Moreover, the analysts' assessment of 60% growth reduction does not square with the fact only 40% of trees (i.e. young trees that were under nine years of age) were deemed susceptible to the moth (MAF 2002b). Given the aforementioned problems, it is not surprising the New Zealand Treasury (in an August 2002 briefing to Cabinet) did not find MAF's upper estimates to be credible, stating that the economic impacts were more likely to be at the lower end of the $58–$356 estimated range (Walsh 2003).

Another problem with MAF's analysis is that most of the estimated economic impacts on the plantation industry were not increased costs brought on by the incursion, but rather estimated profits the sector *might* lose *if* the PAM incursion led to significant defoliation and *if* the defoliation was sufficient to stunt tree growth. Whereas lost profits represented 55% of economic impacts to the plantation industry in a high-impact scenario (i.e., $196.9 million out of $358 million), they represented 100% of economic impacts in the more likely "low-impact" scenario (i.e., $58 million) (MAF 2002b). This has important implications for the estimated total costs to the country. If we exclude the plantation industry's estimated profit losses from the calculation, PAM's predicted 20-year cost would be reduced from $58–$356 million to $13.1–$186.5 million, with the lower figure being five times lower than the eradication operation's final official cost of $65 million (MAF 2002b, 11; Office of the Ombudsman 2007).

Still another problem with the framing is that the economic discussion was one-sided, as it considered the potential economic costs of a PAM incursion without also considering the costs associated with pursuing different biosecurity responses, such as (1) managing the PAM population over time; (2) responding with sterilized moths; or (3) pursuing an aerial pesticide spraying operation. This failure is particularly troublesome when we consider the official final cost of the eradication campaign came in at $65 million (Suckling et al. 2014), which exceeds the government's lower end estimates of what the PAM incursion would cost if they took no action ($58 million) and far exceeds the estimate ($13.1 million) that excludes the forestry industry's potential lost profits. The problem is accentuated when we consider that the $65 million final cost does not consider the eradication campaign's damage to the local ecology, the long-term health costs associated with the residents' exposure to the spray, the health-costs associated with the psychological stresses created by the spraying operation, nor the economic impact of their lost productivity caused by those health problems.

Given these problems, it appears MAF's economic analyses were not honest assessments of likely costs a PAM incursion might incur, nor the costs and benefits associated with each potential response they could have pursued. Rather, it appears their main purpose was to provide a means to stoke people's fears about the moth's potential impact, which would have spurred citizens to support the spraying operation as well as making it more difficult for opposition to emerge towards either eradicating PAM or using pesticides to achieve that eradication.

Framing PAM as a Threat to New Zealand's Ecology

Besides framing PAM as an economic threat, government officials also repeatedly framed it as a threat to native flora.

This occurred prior to the start of spraying. For instance, in December 2001, one month prior to the start of aerial spraying, MAF disseminated a press release that stipulated painted apple moths had been found feeding on indigenous trees, including kowhai, mountain ribbonwood, and karaka. A month later Ruth Frampton (2002) (Director of Forest Biosecurity and the PAM spray operation's manager) reiterated the message in a newspaper op-ed that appeared one week before the start of spraying, specifying that while PAM prefers wattles (a weed) and acacia trees (non-native to New Zealand), they can also feed on three native species (kowhai, karaka, and mountain ribbonwood). That same week she reinforced the point in a Ministry of Primary Industries press release, where she asserted that while aerial spraying was not an ideal solution, it was necessary "to protect New Zealand's forestry and environmental interests" from the moth (MPI 2002).

The messaging was also disseminated at key points during the operation. In September 2002 the Biosecurity minister (Jim Sutton) reiterated the need to protect New Zealand's forestry and environment. It bears noting that this messaging came when he was facing public pressure for having dramatically expanded the program from 928 hectares (2,293 acres) to 7,980 hectares (19,719 acres) (NZ Government 2002a). He reiterated this point three months later, when facing renewed criticism for yet another expansion in the operation, stating "the Cabinet decided to proceed with an eradication attempt because of the incalculable risk the painted apple moth poses to our indigenous forests" (NZ Government 2002b, 1). Similarly, in March 2003, as the government was facing pressure from the release of the Meriel Watts and Hana Blackmore health reports, Judith Tizard (the Member of Parliament representing Central Auckland) argued New Zealand's indigenous forests are threatened by the moth's damage (Tizard 2003, 1).

The argumentation was also used after the spraying operation had ended, in order to defend the operation and MAF. Specifically, four years later, when, following the publication of the Ombudsman's report (which significantly critiqued the PAM operation and MAF), Jim Anderton (the new Biosecurity Minister) stated:

> These moths could have done real damage to our native plants. As a former Aucklander, I know just how important the Waitakere Ranges are to Aucklanders, and Cabinet was well aware of that when we made the decision to try to eradicate the painted apple moth.
>
> (Anderton 2007, 1)

These statements encouraged the public to view the moth as a significant ecosystem menace, which would have bolstered support for calls to eradicate it. Additionally, this frame is likely to have exerted a strong pull on New Zealanders, who pride themselves on having a close relationship to the land (Dew 1999), and even more so on West Aucklanders, who lived close to the Waitakere Ranges Regional Park and many of whom lived in Waitakere City, which was New Zealand's first eco-city.

Problems with the "Ecosystem Threat" Frame

The "ecosystem threat" frame also falters under close scrutiny as evidence suggests it was overstated. Although it is true PAM is able to eat different foods and its spread could have been facilitated by the lack of natural predators, the Island Resource Allocation Hypothesis (Suckling et al. 2014) suggests such invaders have a hard time settling in isolated island floras, such as New Zealand. The reason is that plant species in these settings are less reliant on top–down regulation offered by natural predators and are more protected by "bottom–up" defenses (e.g. phytochemicals or plant architecture) that constrain herbivore reproduction (*Ibid.*). Moreover, Suckling et al. (2014) argue this hypothesis was supported by evidence from PAM's incursion in New Zealand: while the moth was found on three native species, lab studies show that two of those (the karaka and ribbonwood) did *not* support the development of female insects (Suckling et al. 2014).

That is not to say the moth failed to establish itself on plant life in Auckland. The moth was particularly fond of wattles, a very common weed in Auckland, whose numbers were very high in the heavy infestation areas (Suckling et al. 2014). However, wattles is a weed, not part of protected indigenous flora. Besides wattles, numerous moths were found in naturalized acacia and, to a lesser extent, radiata pine. However, both of those species are non-indigenous to New Zealand, not part of the protected indigenous fauna. In fact, some conservationists consider pine trees to be an invasive species that is harmful to indigenous species, and which need to be controlled (Hansford 2021).

Given the deficiencies with the "ecosystem threat" frame and the way that frame was deployed at politically sensitive moments (such as right before the start of spraying, when the spraying was being expanded, and when damning evidence was being released against the spraying operation), it seems that, like the economic threat frame, its use was less an accurate reflection of reality and more a rhetorical ploy to spur public backing of aerial pesticide spraying, in order to protect forestry interests. This assessment is only reinforced when we consider that while neither acacia nor radiata pines are indigenous, both are important to New Zealand's plantation forestry industry, with radiata pine, in particular, representing 90% of plantation forests (MAF 2002b).

Framing PAM as a Threat to Human Health

During the PAM operation government agents used various tactics to portray the moth as a threat to human health and well-being. One tactic was suggesting that exposure to the moth would cause adverse health effects. For instance, in March 2002, two months after the start of the spraying, the Auckland District Health Board (ADHB) released a report alleging "the painted apple moth is known to cause adverse health effects, including skin lesions, eye irritation, and respiratory reactions" (ADHB 2002, 50). As well, in February 2003, at a time when the government was facing political backlash from the Watts and Blackmore health reports, the Ministry for Primary Industries (MPI) put out a media release announcing scientists working with PAM had serious reactions to PAM hairs, and that one experienced "conjunctivitis-type symptoms" when a hair contacted her eye (MPI 2003).

Another tactic was to suggest the health effects were not isolated incidents but rather would affect most people. For example, during a December 10th, 2002, parliamentary debate, when MAF was facing criticism for having significantly expanded the spray zone, Marian Hobbs (Member of Parliament) argued PAM's eradication was important for public health reasons: "scientific evidence is that about 95 percent of people are allergic to the painted apple moth" (NZ Parliamentary Debate 2002). Additionally, a week later Jim Sutton (Director of the Biosecurity government agency) put out a press release stating the PAM eradication campaign was partially driven by the public health risk associated with the moth, arguing that overseas experience had demonstrated that 95% of humans are allergic to PAM hairs (NZ Government 2002a). This message was reiterated in MPI's February 2003 media release, which declared that "just about everyone who has had a reasonable level of exposure to the moths have developed some degree of reaction to the hairs" and that "the impact this moth is having on the PAM project workers will be the same the public would suffer if the moth was allowed to establish here" (*Ibid.*, 1).

A third tactic was to portray the moth as something to be feared. One way this was communicated was by stating that people should seek medical treatment if they come in contact with the moth. An example is MPI's February 2003 press release, which stated "we strongly advise anyone who thinks they have come in contact with a Painted Apple Moth to seek medical attention." The message was also communicated via MAF's use of 30-second radio advertisements that portrayed the moth as scary and dangerous (Derraik 2008). They also communicated this through their "Things get really ugly when you touch the Painted Apple Moth" advertisement, which ran numerous times in Auckland's main newspaper (i.e. *The New Zealand Herald*). While the ad's title, in and of itself, suggests the moth is something to be avoided, if not feared, the ad copy bolsters the message by stating "should you touch it (and please don't) it can give you a nasty reaction. That's because the caterpillar sheds its hairs which cause itchy rashes, skin lesions, eye irritations and significant distress" and "The health hazard, however, doesn't stop there. People exposed to the caterpillar are likely to experience a reaction" (MAF 2002f). Regarding the latter statement, the vagueness of what that "reaction" could be only amplifies the fear that could be conjured by the messaging. Prompting further anxiety, particularly among parents, was the ad's statement that "Children are particularly susceptible."

In 2007, three years after the spraying campaign had ended, Jim Anderton (new Biosecurity Minister) (2007, 1) reiterated the frame, stating that PAM's "caterpillar has a very toxic effect on human beings" and that skin contact provokes serious toxic reactions. Moreover, he emphasized that the spraying campaign was partially driven by concerns for public health, including the "extreme allergic reactions most people tend to have to the caterpillars of the moth" (*Ibid.*).

Fourth, government communications suggested that if PAM became established, it would curtail the easy access to nature New Zealanders know and love. For instance, the MPI's February 2003 media release stated, "an infestation of Painted Apple Moth in our native forests could cause authorities to close areas to the public during the height of the moth breeding season because of the impact it could have on public health" (MPI 2003).

As a whole, these communications signaled to the public that PAM posed a significant threat to New Zealanders' health and lifestyle. In turn, this frame would have also contributed to drumming up and maintaining support for the government's eradication efforts, which would have made it more difficult to build successful opposition to the aerial spraying operation.

Problems with the "Threat to Public Health" Frame

As with the previous two frames, there were significant problems with framing the moth as a human health menace. First, there wasn't persuasive empirical support to support the claim that PAM had human health impacts. For instance, the 2002 ADHB report failed to provide any empirical evidence for its claim that PAM causes adverse health effects. This is surprising given that the moth is a common pest in suburban orchards and urban gardens throughout South Australia, including most of Tasmania (ADHB 2002; Derraik 2008). As Derraik (2008) points out, "if the claims of widespread human susceptibility and consequent adverse reactions were indeed accurate, one would expect case reports of human exposure in Australia" (37).

Instead of relying on actual data, the ADHB's assertions were based on the reported health effects associated with two other members of the moth family: the tussock moth (*Lymantriidae*) and the gypsy moth (*Lymantria dispar dispar*). While they belong to the same moth family, there seems to be no other rationale for using them as a basis for evaluating the potential health impact of the PAM (Derraik 2008). In particular, Derraik (2008) emphasized there was no evidence their biochemical profiles are similar enough to cause equivalent reactions in humans. Additionally, while the scientific literature provides extensive evidence about the health effects of other moths, very little is reported about PAM and what is reported suggests any health effects would be minor (*Ibid.*).

There were also problems with the claim that nearly everyone exposed to painted apple moths would experience health effects. While Marian Hobbs and Jim Sutton asserted 95% of humans are allergic to PAM, neither provided any support for their claims. Similarly, MAF failed to substantiate its claim that multiple lab technicians had suffered health effects. Nor did they substantiate the claim that one lab technician had suffered conjunctivitis-type symptoms. Additionally, even if that case was substantiated, one case is hardly sufficient to generalize the problem to 95% of the human population, particularly when we consider that, unlike lab technicians, 99.9% of citizens would not have spent hours every day working closely with the moths.

As for the possibility that a PAM infestation might close off forests to the public, this too seems to have been an overblown claim. During the PAM operation Gordon Hosking (Director of the 1996–1997 Project Ever Green to exterminate the white-spotted tussock moth) asserted that forest closures due to infestation are exceedingly rare overseas and that this even holds for large outbreaks (Hosking as cited in Walsh 2003). Moreover, he emphasized that beneficiaries of eradication operations tend to overestimate the dangers posed by pests (*Ibid.*).

Beyond deficiencies with the government's "threat to human health" frame, there is reason to believe the government's stated concern about human health was disingenuous, as New Zealand governments have had, and continue to have, a long track record of repeatedly placing citizens in harm's way. For instance, during the 1960s, 1970s, and 1980s, the government allowed the spraying of 2,4,5-T throughout New Zealand. 2,4,5-T was one of the two herbicides making up Agent Orange, the powerful defoliant that the U.S. military used against Northern Vietnamese during the Vietnam War. The herbicide was laced with carcinogenic dioxins and associated with severe skin rashes, malaise, peripheral nervous system disturbances, liver toxicity, cancer, birth defects, and miscarriages (Bunting 2013; Dew 1999; Wildblood-Crawford 2008). Aside from allowing that toxicant to be sprayed on New Zealand soil, between 1981 and 1985 the government actually subsidized farmers to use the herbicide, resulting in 53,795,834 liters being sprayed over New Zealand in 1985 alone (Takoko & Gibbs 2004). In a second example, for 40 years the government allowed the logging industry to use pentachlorophenol (PCP) as a wood preservative for their radiata pine exports (Dew 1999). Besides being associated with a host of health effects (including fever, fatigue, weight loss, nausea, and mood swings), the PCP grades used in New Zealand contained toxic impurities, including dioxins and furans, that were associated with their own set of health problems (*Ibid.*). The practice went on for 40 years before a government inquiry was finally called in 1991 (*Ibid.*). A third example is the 1996–1997 urban aerial spraying campaign to eliminate the white-spotted tussock moth from the Auckland region, which also relied on the Foray 48B pesticide and which led 375 residents to complain about health effects (Ombudsman 2007). Fourth, as of 2004 the New Zealand government was aware of 7,000 sites around the country that were potentially contaminated by hazardous substances but were withholding the information from the public, in order to protect property values and developers (Gardiner & Gamble 2004).

While there is no doubt some government agencies (such as the Ministry of Health) carry out activities to protect human health, the above examples demonstrate many instances where other government agencies have allowed citizens to be repeatedly placed in harm's way. Moreover, as is discussed in Chapter 8, the Ministry of Health has been known to abdicate its responsibility to protect public health when it conflicts with the priorities of other government agencies, such as the Ministry for Agriculture and Forestry. This points to the fact that in New Zealand, at least, governments tend to prioritize economic productivity over public health concerns.

Ideological Factors that Mediated the Effectiveness of Government Framing Activities

Thus far, we've seen that government officials pursued considerable efforts to frame PAM as a biosecurity threat. Moreover, there is evidence to suggest their efforts helped build support for both eradication and aerial pesticide spraying, which, conversely, would have slowed the spread of opposition to spraying activities.

However, it bears noting that such successes do not occur in an ideological vacuum, as citizens possess pre-existing belief systems that will mediate their propensity to accept the government's framing. For example, if a community tends to believe that foreign species are destructive to local ecosystems, then they will be primed to accept government attempts to simplistically frame a new species as the source of ecosystem problems. On the other hand, if the citizenry understands that a foreign species' impact will be determined by the ecosystem's health, they are likely to resist attempts to simplistically frame a species as the source of the problem. Additionally, they are likely to call for a more comprehensive and more nuanced understanding of the problem, which would include understanding how human actions have destabilized local ecosystems and made them more vulnerable to the invasive species in question. Thus, it is crucial to analyze and understand the aspects of the cultural context that will mediate the effectiveness of the government's communication activities.

Towards that end, the rest of this chapter examines three factors that primed New Zealanders, and Aucklanders in particular, to buy into the government's framing of PAM as a biosecurity threat: (1) a long history of struggling with invasive species; (2) a weak eco-literacy; and (3) the important place the environment holds in the country's national identity.

New Zealand's History of Struggling with Invasive Species

A factor that predisposed the population to accept the government's frame was the country's long and well-documented history of struggling with invasive species, including rabbits, deer, certain birds, stouts, ferrets, invasive plants, insects, and possums (Barker 2008; Green 2000; Isern 2002; Peden 2008). Not only have such species successfully established a foothold in New Zealand, many have also caused substantial ecological and economic damage. For instance, New Zealand's population of 60–70 million possums have had a devastating effect on tree growth as they are estimated to consume 21 tons of vegetation per night, most of which is new growth (Isern 2002; Richie 2000). They have also outcompeted local fauna for food resources and shelter (Clout 2006). As well, they have been a constant threat to some indigenous species, consuming eggs, birds, and even bats (Brockie 2015). Possums are also a problem for farmers because they eat agricultural produce and spread tuberculosis to farm animals (Ritchie 2000). These problems are estimated to cost farmers $35 million a year, which has compelled the New Zealand government to spend vast sums on possum control (an estimated $80 to $110 million annually (Predator Free NZ n.d.; Warburton et al. 2009). The country's history of struggling against invasives has predisposed New Zealanders to view invasive species as potential threats, which would have primed them to more easily accept the government's vilification of PAM.

Besides a history of struggling with foreign species, the citizens' concerns about them would have been reinforced in the years leading up to the PAM incursion, due to the well-publicized contemporary incursions of numerous other species. One was the Argentine ant (*Linepithema humile*), which arrived in 1990, following the Commonwealth Games that were held in Auckland that year and spread throughout the country (PCE 2000). Another was the 1996–1997 incursion of the white-spotted tussock moth (*Orgyia thyellina*), which, like PAM, resulted in aerial pesticide spraying of Auckland and which, as a result, was extensively covered by the mass media (Hosking et al. 2003). A third was the arrival of the varroa mite (*Varroa destructor*), which attacks and feeds on honey bees. The mite arrived in 2000 and media reported it would cost the apiculture industry $600 to $900 million over the subsequent 35 years (NZ Herald 2001). Still another case was the Asian tiger mosquito (*Aedes albopictus*), which can be a serious public health hazard as it can be the vector for yellow fever, dengue fever, Ross River virus, West Nile virus and chikungunya virus (Derraik 2004). This mosquito was detected by government officials on nine different occasions between 1993 and 2001 (*Ibid.*). Adding to the media drumbeat against invasive species, in 1998 the Ministry of Health began disseminating media releases about mosquito detections (*Ibid.*). Beyond the communications about those specific cases, in 1997 the New Zealand Parliamentary Commissioner released the *State of New Zealand's Environment* report (Taylor 1997), which detailed the country's rate of biodiversity loss and partially attributed it to invasive species.

This series of events would have helped intensify the citizenry's concerns about invasive species, which would have increased their propensity to accept the government's framing of PAM as an ecological threat.

The Population's Shallow Ecological Literacy

A second contributing background factor was the citizenry's lack of a deep ecological literacy. Ecological literacy refers to an understanding of (1) how ecosystems work; (2) how a mismanagement of an ecosystem could lead to the proliferation of unwanted species; and (3) the kinds of non-toxic interventions that could be pursued to re-balance the ecosystem in question and reduce, if not eliminate, the problematic species (Coppolla 1999). When the citizenry possesses this knowledge they are more likely to resist government attempts to convince them that one species is the root cause of their ecosystem problems or that pesticides are the best solution to resolve such problems. Conversely, when the citizenry lacks a deep eco-literacy, it makes it easier for government agencies to promulgate a superficial understanding of the ecological problem and to sell pesticides as the best, and often only, way to resolve the ecosystem issues.

Unfortunately, citizens in most First World nations lack an adequate eco-literacy and an important reason is that education systems fail to provide it. This is particularly true of high education, as has been noted by David Orr (1994) and many others (Cortese 2003; Haigh 2005; M'Gonigle & Starke 2006; Wolfe 2001). It is true that since the 1970s universities have significantly expanded the number of courses and degrees offered on environmental topics (Brint et al. 2009; Collett & Karakashian 1996; Johnson et al. 2020). However, a fundamental limitation of those efforts is that these courses are invariably electives. Previous research on the issue in the United States found that less than 11% of tertiary institutions require students to complete a course that will increase their environmental awareness, with public institutions having an even lower rate of 7% (Wolfe 2001). A similar situation has existed in New Zealand, where, as of 2022, an environmental awareness graduation requirement is still missing at all seven universities. In turn, the population's lack of deep eco-literacy would have further predisposed them to buy into the government's vilification of PAM.

The Environment's Importance in New Zealand's National Identity

Framing PAM as a threat to local eco-systems would have struck a strong chord with New Zealanders, who perceive themselves as having a strong relationship with the environment. As discussed by Kevin Dew (1999), New Zealanders have long prided themselves on their close relationship to the environment, which has helped ground their national identity. While this grounding was originally based on the pioneering spirit associated with taming a wild land (Bell 1996; Phillips 1996), in the second half of the twentieth century it became increasingly associated with outdoor activities, such as hiking, spending time at the beach, sailing, surfing, and other ocean activities. Moreover, in the 1960s New Zealanders began to become more knowledgeable about the environmental havoc that a century of settlement had wrought on the land, which has included the draining of the country's bountiful wetlands, vastly diminished forests, polluted rivers, and reduced biodiversity. The growing awareness of environmental destruction contributed to the New Zealand environmental movement's emergence in the early 1970s and reinforced the environment's place in New Zealanders' sense of identity.

That sense of identity was further burnished by the environmental movement's many subsequent successes, which included the campaign to save the Manapouri Valley from being flooded to establish a South Island hydroelectric operation. This campaign garnered widespread support, which included getting 10% of eligible voters to sign a petition opposing the project, which led the political establishment to distance itself from the project and to eventually abandon it (Knight 2018). Another environmental success was making New Zealand a nuclear-free country through the 1987 *New Zealand Nuclear Free Zone, Disarmament and Arms Control Act* (*Ibid.*). Beyond rejecting the installation of nuclear energy production, the policy prohibited nuclear-powered warships from docking in New Zealand waters. This drew the enmity of the United States, which retaliated in numerous ways, including classifying New Zealand as a "former ally" for well over a decade (Huntley 1996).

New Zealand's environmental identity was further bolstered in the mid-1990s, when Jenny Shipley's government sought to cash in on the country's

As well, the analysis underscores how using the "biosecurity threat" discourse can serve as a powerful controlling process. Not only can it manipulate people through fear, but it can also stifle dissent, which has chilling implications for democracy. Consequently, when this discourse is deployed, analysts should carefully scrutinize the effects of its deployment as well as ask who is and is not served by its deployment.

This chapter also illuminated cultural factors that predisposed New Zealanders to accept the government's framing of the moth. This signals that government "naturework" does not occur in a cultural vacuum and that it is important to assess the cultural landscape within which those activities are being pursued. Doing so will help us identify and understand the ideological factors that mediate citizen receptiveness to such messaging.

Beyond its relevance for understanding urban aerial pesticide spraying operations, the insights are also germane for understanding the ideological work pursued to build support for other pesticide uses in urban contexts, such as weed control on sports fields, golf courses, roadways, and parks. In each case potent social forces (including government agencies, corporations, their public relations henchmen, and a complicit mass media) will carry out ideological work to persuade the public to view the targeted species as a significant threat, and the effectiveness of such naturework will be mediated by the surrounding cultural context.

Having said this, portraying a foreign species as a threat is only one part of getting the public to support a pesticide spraying operation. The other consists of allaying the health concerns that citizens might have about the pesticide. Illuminating that process is what the next two chapters will address.

References

Anderton, Jim. 2007. "Painted Apple Moth spray campaign was vital." *Beehive.govt. nz*, November 12, 2007. Accessed December 15, 2017. http://www.beehive.govt. nz/node/31259

Auckland District Health Board, Public Health Service. 2002. *Health Risk Assessment of the 2002 Aerial Spray Eradication Programme for the Painted Apple Moth in Some Western Suburbs of Auckland*. Auckland District Health Board.

Barker, Kezia. 2008. "Flexible Boundaries in Biosecurity: Accommodating Gorse in *Aotearoa* New Zealand." *Environment and Planning A* 40: 1598–1614.

Bell, Claudia. 1996. *Inventing New Zealand: Everyday Myths of Pakeha Identity*. Auckland, NZ: Penguin Press.

Beston, Anne. 2001. "Big Spray Targets Western Suburbs." *New Zealand Herald*, November 9.

Blackmore, Hana. 2003. *Interim Report of the Community-based Health and Incident Monitoring of the Aerial Spray Programme*. Auckland, New Zealand.

Brint, Steven, and Lori Turk-Bicakci, Kristopher Proctor, and Scott Patrick Murphy. 2009. "Expanding the Social Frame of Knowledge: Interdisciplinary Degree-granting fields in American Colleges and Universities, 1975-2000." *The Review of Higher Education* 32(2): 155–183.

Brockie, Bob. 2015. "Introduced animal pests – Possums." *Te Ara – The Encyclopedia of New Zealand.* https://teara.govt.nz/en/introduced-animal-pests/page-2

Bunting, Zoe. 2013. *Subsidising Agricultural Expansion at the Expense of Public Health: The Case of 2,4,5-T and the Role of the New Zealand Government.* Unpublished MA thesis, Department of Sociology, University of Auckland.

Capek, Stella. 2009. "The Social Construction of Nature: Of Computers, Butterflies, Dogs, and Trucks." In *Twenty Lessons in Environmental Sociology,* edited by K. Gould and T. Lewis, 11–24. New York: Oxford University Press.

Cockburn, Andrew. 2015. "Weed Whackers: Monsanto, Glyphosate, and the War on Invasive Species." *Harper's Magazine,* September: 57–63.

Collett, Jonathan and Stephen Karakashian. 1996. "Turning curricula green." *The Chronicle of Higher Education* 42(24): B1–B2.

Coppola, Nancy. 1999. "Greening the technological curriculum: a model for environmental literacy." *The Journal of Technology Studies* 25(2): 39–46.

Cortese, Anthony. 2003. "The Critical Role of Higher Education in Creating a Sustainable Future." *Planning for Higher Education* 31(3): 15–22.

Clout, Mick N. 2006. "Keystone Aliens? The Multiple Impacts of Brushtail Possums." In *Biological Invasions in New Zealand,* edited by Robert B. Allen and William G. Lee, 265–279. Berlin, Germany: Springer Press.

Derraik, José G. B. 2004. "Exotic Mosquitoes in New Zealand: A Review of Species Intercepted, Their Pathways and Ports of Entry." *Australian and New Zealand Journal of Public Health* 28(5): 433–444.

Derraik, José. 2008. "The Potential Direct Impacts on Human Health Resulting from the Establishment of the Painted Apple Moth (*Teia anartoides*) in New Zealand." *The New Zealand Medical Journal* 121(1278): 35–40.

Dew, Kevin. 1999. "National Identity and Controversy: New Zealand's Clean Green Image and Pentachlorophenol." *Health and Place* 5: 45–57.

Fine, Gary Alan. 1988. *Morel Tales: The Culture of Mushrooming.* Cambridge, MA: Cambridge University Press.

Forest Research. 1999. *Forest Health News* 85, May 1999.

Frampton, Ruth. 2002. "Dialogue: Well-considered approach to swatting indiscriminate pest." *New Zealand Herald,* January 14, 2002. Accessed September 30, 2015. http://www.nzherald.co.nz/nz/news/article.cfm?c_id=1&objectid=586604

Gardiner, James and Warren Gamble. 2004. "Secrecy keeps lid on toxic cesspits." *New Zealand Herald,* September 9. Accessed October 15, 2015. http://www.nzherald.co.nz/nz/news/article.cfm?c_id=1&objectid=173615

Green, Wren. 2000. *Biosecurity Threats to Indigenous Biodiversity in New Zealand – An Analysis of Key Issues and Future Options.* A Background report prepared for the Parliamentary Commissioner for the Environment, Wellington.

Haigh, Martin. 2005. "Greening the University Curriculum: Appraising an International Movement." *Journal of Geography in Higher Education* 29(1): 31–48.

Hansford, Dave. 2021. "The March of the Pines." *New Zealand Geographic* 171: 88–105.

Hosking Gordon, John Clearwater, John Handiside, Malcolm Kay, J. Ray and N. Simmons. 2003. "Tussock moth eradication - a success story from New Zealand." *International Journal of Pest Management* 49(1): 17–24.

Huntley, Wade. 1996. "The Kiwi that Roared: Nuclear-free New Zealand in a Nuclear-Armed World." *The Nonproliferation Review* 4(1): 1–16.

Isern, Thomas. 2002. "Companions, Stowaways, Imperialists, Invaders: Pests and Weeds in New Zealand." In *Environmental Histories of New Zealand*, edited by Eric Pawson and Tom Brooking, 233–245. Melbourne, AU: Oxford University Press.

Johnson, Erik, Ali. Ilhan and Scott Frickel. 2020. "Riding a long green wave: interdisciplinary sciences and studies in higher education." *Environmental Sociology* 6(4): 433–448.

Jones, Geoffrey and Simon Mowatt. 2016. "National image as a competitive disadvantage: the case of the New Zealand organic food industry." *Business History* 58(8): 1262–1288.

Joy, Mike. 2015. *Polluted Inheritance: New Zealand's freshwater crisis*. Wellington, New Zeland: Bridget Williams Books.

Knight, Catherine. 2018. *Beyond Manapouri: 50 Years of Environmental Politics in New Zealand*. Christchurch, New Zealand: Canterbury University Press.

M'Gonigle, Michael and Justine Starke. 2006. *Planet U: Sustaining the World, Reinventing the University*. Gabriola Island, B.C.: New Society Publishers.

Ministry of Agriculture and Forestry (MAF). 2000. *Potential Economic Impact on New Zealand of the Painted Apple Moth*, July 2000, MAF Policy.

Ministry of Agriculture and Forestry (MAF). 2001. "New Plans for Target Aerial Spraying." *Scoop Independent News*, December 13, 2001. https://www.scoop.co.nz/stories/AK0112/S00016.htm

Ministry of Agriculture and Forestry (MAF). 2002a. "Spray One against Moth Completed." *Scoop Independent News*, January 24, 2002. https://www.scoop.co.nz/stories/GE0201/S00029.htm

Ministry of Agriculture and Forestry (MAF). 2002b. *Painted Apple Moth: Reassessment of potential economic impacts*, May 7, 2002. Accessed November 9, 2015. http://www.biosecurity.govt.nz/files/pests/painted-apple-moth/pam-reassessment-economic-impacts.pdf

Ministry of Agriculture and Forestry (MAF). 2002c. "MAF Plans to Abandon Painted Apple Moth Project." *Scoop Independent News*, June 24. https://www.scoop.co.nz/stories/PO0206/S00156/maf-plans-to-abandon-painted-apple-moth-project.htm?from-mobile=bottom-link-01

Ministry of Agriculture and Forestry (MAF). 2002d. "MAF response to PAM decision." *Scoop Independent News*. September 10, 2002. https://www.scoop.co.nz/stories/PO0209/S00039.htm

Ministry of Agriculture and Forestry (MAF). 2002e. "Painted Apple Moth Eradication Project Fact Sheet." *Scoop Independent News*, September 10, 2002. https://www.scoop.co.nz/stories/AK0209/S00024.htm

Ministry of Agriculture and Forestry (MAF). 2002f. "Things get really ugly when you touch the Painted Apple Moth." *New Zealand Herald*, October 10, 2002. p. A10.

Ministry for Primary Industries (MPI). 2002. "Response to Meriel Watts article MAF bungles the biosecurity in West Auckland." *MPI.govt.nz*, January 14. Accessed January 22, 2015. http://www.mpi.govt.nz/news-resources/news/response-to-meriel-watts-article-maf-bungles-the-biosecurity-in-West-Auckland.aspx

Ministry for Primary Industries (MPI). 2003. "Technician Removed from Painted Apple Moth Rearing Project after Reaction to Moth Hairs." *Ministry of Primary Industries media release*, February 03, 2003. http://www.mpi.govt.nz/news-resources/news/technician-removed-from-painted-apple-moth-rearing.aspx

Nader, Laura. 1997. "Controlling Processes: Tracing the Dynamic Components of Power." *Current Anthropology* 38(5): 711–738.

New Zealand Government. 2002a. "Cabinet decides on eradication for PAM." *Scoop Independent News,* September 9, 2002. Accessed April 14, 2020. https://www.scoop.co.nz/stories/PA0209/S00133.htm

New Zealand Government. 2002b. "All New Zealanders need to work together for effective biosecurity." *Beehive.govt.nz*, December 17, 2002. Accessed October 11, 2015. http://www.beehive.govt.nz/release/all-new-zealanders-need-work-together-effective-biosecurity

New Zealand Government. 2004. "Painted Apple Moth spray programme ends." *Beehive.govt.nz*, May 13, 2004. Accessed October 17, 2015. http://www.beehive.govt.nz/release/painted-apple-moth-spray-programme-ends

New Zealand Herald. 2001. "Biosecurity: Common questions." *New Zealand Herald,* October 18. https://www.nzherald.co.nz/nz/biosecurity-common-questions/2KQET5STCMWQTMAENXQOP36PBQ/

New Zealand Parliamentary Debate. 2002. *Questions for Oral Answer Questions to Ministers: Industry and Regional Development, Painted Apple Moth---Vegetation Control Zone, West Auckland.* December 10, 2002. Accessed December 14, 2013. http://www.vdig.net/hansard/archive.jsp?y=2002&m=12&d=10&o=9&p=9

Office of the Ombudsman (Mel Smith). 2007. *Report of the Opinion of Ombudsman Mel Smith on Complaints Arising from Aerial Spraying of the Biological Insecticide Foray 48b on the Population of Parts of Auckland and Hamilton to Destroy Incursions of Painted Apple Moths, and Asian Gypsy Moths, Respectively.* Wellington, New Zealand: Office of the Ombudsman.

Organisation for Economic Co-operation and Development (OECD). 2007. *OECD Environmental Performance Reviews New Zealand.* Paris: OECD.

Orr, David. 1994. "Education's challenge: recalibrating values." *Forum for Applied Research and Public Policy* Fall: 48–50.

Painted Apple Moth Community Advisory Group (PAM CAG). 2002. "Re: Moth Pest Spray Date Announced." *Scoop Independent News,* July 4. https://www.scoop.co.nz/stories/PO0207/S00055.htm

Parliamentary Commissioner for the Environment. 2000. *New Zealand under Siege: A Review of the Management of Biosecurity Risks to the Environment.* Wellington, NZ: Office of the Parliamentary Commissioner for the Environment.

Peden, Robert. 2008. "Rabbits." *Te Ara – The Encyclopedia of New Zealand.* https://teara.govt.nz/en/rabbits/print

Phillips, Jock. 1996. *A Man's Country? The Image of the Pakeha Male: A History.* Auckland, New Zealand: Penguin Press.

Predator Free NZ. n.d. "Possum Facts." https://predatorfreenz.org/resources/introduced-predator-facts/possum-facts/

Richie, J. 2000. *Possum: Everybody's Problem.* Department of Conservation, Wellington, N.Z.

Royal Forest and Bird Protection Society. 2001. "Spread of moth is a biosecurity emergency." *Scoop Independent News,* December 3. https://www.scoop.co.nz/stories/AK0112/S00006.htm

Safe Waitekere. 2006. *Safe Community Re-Designation Application.* https://isccc.global/files/custom/Community/46waitakere_re.pdf

Simon, Nathan. 2009. "Conservation – A History – Environmental Activism, 1966–1987". *Te Ara Encyclopedia of New Zealand.* https://teara.govt.nz/en/conservation-a-history/page-8

Suckling, David, John Charles, Malcolm Kay, John Kean, Graham Burnip, Asha Chhagan, Alasdair Noble, and Anne Barrington. 2014. "Host Range Testing for Risk Assessment of a Sexually Dimorphic Polyphagous Invader, Painted Apple Moth." *Agricultural and Forest Entomology* 16: 1–13.

Sutton, Jim. 2004. "Painted Apple Moth spray programme ends." *Beehive.govt.nz - The official website of the New Zealand Government.* Accessed: July 15, 2022. https://www.beehive.govt.nz/release/painted-apple-moth-spray-programme-ends

Takoko, Mere and Andrew Gibbs. 2004. *Nga matitapu o te hakino - People poisoned daily: How Ivon Watkins Dow Contaminated Aotearoa.* Auckland: Greenpeace NZ. https://www.dioxinnz.com/pdf-wai262/Section%20G.pdf

Taylor, Rowan. 1997. *The State of New Zealand's Environment: 1997.* Wellington, New Zealand: The Ministry for the Environment.

Tizard, Judith. 2003. "Painted Apple Moth eradication campaign necessary." *Beehive.govt.nz,* March 29, 2003. https://www.beehive.govt.nz/release/painted-apple-moth-eradication-campaign-necessary

Tyson, Janet. 2006. "The Painted Apple Moth Programme (C)." The Australian and New School of Government.

Walsh, Frances. 2003. "Wipe Out." *Metro Magazine* March 2003: 44–51.

Warburton, Bruce, Phil E. Cowan, and James Shepherd. 2009. "How Many Possums Are Now in New Zealand Following Control and How Many Would There Be Without It?" Prepared for Northland Regional Council. *Landcare Research – Manaaki Whenua.* https://envirolink.govt.nz/assets/Envirolink/720-NLRC104-Possum-numbers-inNZ.pdf

West Aucklanders Against Aerial Spraying (WASP). 2002. "Helen Wiseman-Dare Speech: Anti Spray March Nov 30." *Scoop Independent News,* December 4.

Wildblood-Crawford, Bruce S. 2008. *Environmental (In)Justice and 'Expert Knowledge': The Discursive Construction of Dioxins, 2,4,5-T and Human Health in New Zealand, 1940 to 2007.* Unpublished PhD Thesis, Department of Geography, University of Canterbury.

Wolfe, Vicki. 2001. "A survey of the environmental education of students in non-environmental majors at four year institutions in the USA." *International Journal of Sustainability in Higher Education* 2(4): 301–315.

7 Government Actions that Allay Pesticide Concerns

While framing a foreign species as a biosecurity threat can build public support for pesticide spraying, that is only part of the equation. The other part consists of allaying concerns about the pesticide that will be used, as previous research has shown that safety is a crucial mediator of public support for pesticide use (Coppin et al. 2002; Dunlap & Beus 1992). Regarding the PAM spraying operation, government officials pursued several activities that allayed Aucklanders' concerns about the spraying operation, and this chapter covers three: (1) proposing an operation at the outset that was much smaller in scope and duration than would eventually get deployed; (2) incremental expansion of the spraying operation; and (3) portraying the Foray 48B pesticide and its ingredients as being harmless to humans.

Proposing a Small, Limited, and Less Threatening Initial Plan

At the outset the Ministry of Agriculture and Forestry (MAF) proposed a spraying operation that was far smaller and far less threatening than what it would eventually become. This, in and of itself, would have served to tone down Aucklander concerns and minimize the spread of opposition.

One aspect that grew over time was the size of the spray zone, as the area MAF initially proposed to spray was only a fraction of what it would eventually become. Specifically, at 300 hectares (741 acres), the proposed spray zone was only 1/35th the size of the 10,632 hectare (26,272 acre) spray zone used a year later (Goven et al. 2007). Perhaps more importantly, it was noticeably smaller than the spray zones used for the 1996–1997 Project Ever Green, which West Aucklanders may have been vaguely familiar with. Specifically, the PAM operation's original 300 hectare spray zone was 1/3rd the size of the smallest spray zone for the 1996–1997 operation [which was 1,000 hectares (2,471 acres)] and 1/13th the size of that operation's peak spray zone [i.e. 4,000 hectares (9,884 acres)] (Office of the Ombudsman 2007). Even though the PAM spray zone doubled in size right before the start of the spraying, that was still considerably smaller than either the 1996–1997 spraying zones or what the PAM spray zone would eventually become.

DOI: 10.4324/9780429426414-8

Another operation aspect that changed was the spray zone's per capita density. The original zone's density was quite low as it included thinly populated riparian areas around the Waikumete Cemetary and along the Whau River estuary (Office of the Ombudsman 2007). As the spray zone expanded, it increasingly contained densely populated neighborhoods (MAF 2001d). In turn, these changes significantly increased the number of households that were exposed to the pesticide. Where MAF initially predicted 600 residential and 200 industrial properties would be exposed to the spraying, at the operation's peak 43,000 homes were exposed to the spraying (Beston 2001; Office of the Ombudsman 2007). With an increase in affected properties came an increase in the number of people exposed to the spraying. Where there were approximately 13,500 people in the original spray zone, it is estimated there were over 200,000 people in the largest spray zone (Blackmore 2020; Office of the Ombudsman 2007).

Initially proposing a smaller and less threatening operation benefitted MAF as it minimized the number of residents who felt threatened at the outset and who would have felt compelled to agitate against the spraying operation. If all 200,000+ people who were exposed during the operation's peak understood at the outset that they would eventually be exposed to and potentially impacted by the spraying, a significant number would have become concerned, if not alarmed, at the start of the spraying operation. In turn, substantial numbers would have joined the ranks of the initial opposition, which, one imagines, would have considerably accelerated the spread of opposition efforts.

Another operational detail that significantly changed was the number of required sprayings. Where the public was originally told the operation would consist of six to eight sprayings, by the end MAF had conducted aerial sprayings on more than 60 days, with double sprayings (i.e. airplane sprayings were followed up by helicopter sprayings to reach more inaccessible areas) occurring on many of those days (Goven et al. 2007; MAF 2001e; Office of the Ombudsman 2007). With an increase in sprayings came an increase in the operation's duration, which increased from the four-month timeframe that was initially proposed to over 29 months (MAF 2001e; Office of the Ombudsman 2007). As was the case with other changes to the operation, if residents initially understood the full number of sprayings they would have to contend with, many would have resisted the operation when it was being proposed. The same is true for the timeframe expansion. It is one thing to put up with a spraying operation that lasts four months but quite another to put up with a spray operation that lasts nearly two and a half years.

Significant changes also occurred with the pesticide delivery. Where the original proposal was to use helicopters to spray inaccessible "hot spots" in a targeted fashion, after four months of spraying MAF began to increasingly rely on airplanes to deliver the pesticide (MAF 2001d, 2002b; Office of the Ombudsman 2007). This considerably increased the spray drift that residents were already having to contend with, which increased the impact on gardens (organic gardeners lost their certification when their crops were exposed to the pesticide), cars (which got covered with a sticky residue), and homes (where the pesticide often lingered in the air for hours after the spraying) (New Zealand Herald 2003; WASP 2002; Teschke et al. 2001). This operational change also mattered for perceptions of pesticide safety and the spread of opposition. If residents initially knew the pesticide would eventually be delivered by planes and would cause spray drift, many would have dropped their support for the operation, with many turning to campaigning against the operation. This would have been particularly true for those who understood how spray drift would impact their personal lives.

There is the question of whether the government deliberately misled the public about the need for a spraying operation. It seems the initial problem wasn't deception but rather technical incompetence. During the eradication operation's first three years (i.e. 1999–2002), operation leaders made one mistake after another, which included repeatedly failing to accurately assess the size of the incursion, the speed of its spread, and what interventions were needed to effectively contain and then eradicate the Painted Apple Moth (OAG 2002; Panckhurst 2001; Walsh 2003). This provoked a string of changes, which suggested leadership was in over their heads. This point is underscored by the fact opposition groups repeatedly demanded Ruth Frampton's (the original PAM operation director) resignation, which led her to eventually step down in May 2002 (Green Party 2002a; PAM CAG 2001).

On the other hand, government officials also proved deceitful on many occasions, which included its portrayal of PAM as a triple biosecurity threat (see Chapter 6) and its portrayal of Foray 48B pesticide as being harmless to humans (see below). Additionally, in a November 2001 press release entitled "No Blanket Spraying" David Cunliffe (the Labour Party Member of Parliament for West Auckland) stated that MAF had no plans to blanket spray large areas in West Auckland and that if the initial operation was unsuccessful by May 2002, government would extensively consult the public on how to proceed (New Zealand Labour Party 2001). As we now know, MAF did eventually resort to blanket spraying and summarily ignored public input about the issue (Green Party 2002b). Another example of deception occurred prior to the July 2002 election, when MAF publicly opposed continuing the spraying operation, only to reverse course and substantially expand the operation *after* the Labour government was re-elected that month (Office of the Ombudsman 2007).

Regardless of whether it was done intentionally, as a result of incompetence or due to negligence, keeping citizens in the dark about the operation's eventual scale served as an effective controlling process, as it minimized both the number of citizens who might become alarmed about the proposed operation and the number who might agitate against it at the outset. In turn, minimizing these numbers reduced the chances that strong opposition would emerge at the beginning, when the operation still needed to win public acceptance and was still, therefore, vulnerable. Minimizing opposition at the start allowed MAF to get the operation underway, after which it gathered institutional momentum that made it much harder to effectively derail in the weeks and months that followed.

Incrementally Expanding the Spraying Operation

A controlling process Laura Nader (1997) writes about is "cumulative tinkering," where cultural changes are made in small increments over time, as opposed to being made all at once. She asserts that "cultural control is often the result of incremental, not abrupt, change, and when it is achieved it is powerful indeed because it slides in rather unnoticed and comes to be considered natural" (Nader 1997: 722).

As this pertains to the PAM operation, there were numerous aspects of the operation that grew incrementally over time, with an important one being the spray zone. While the spray area was originally set at 300 hectares (741 acres) in October 2001 (when the aerial spraying operation was first proposed), in December 2001 (a month before the start of the spraying) government officials expanded it to 626 hectares (1,547 acres), after which they expanded it to 722 hectares (1,784 acres) in July 2002 (Beston 2001; Goven et al. 2007; MAF 2001e, 2002b). Then, after the Labour government secured re-election in July, MAF further expanded it to 868 hectares (2,145 acres) in August and 962 hectares (2,377 acres) in September (MAF 2002c, 2002d). Then MAF announced there would be a significant increase to 7,980 hectares (19,719 acres) in October, after which it was incrementally increased to 8,686 hectares (21,464 acres) on December 2nd, 8,903 hectares (22,000 acres) on December 20th, and in January reached its peak coverage of 10,632 hectares (26,272 acres) (Goven et al. 2007; MAF 2002e, 2002f, 2002h; Office of the Ombudsman 2007).

Correspondingly, an incremental expansion in the spray zone meant a corresponding incremental expansion in the number of properties being affected. Where 800 were in the spray zone that was originally proposed, when spraying began (in January 2002) the number had expanded to 3,000 properties, with the number expanding further to 4,000 in September 2002, over 37,500 in early October, and then 40,000 in late October (Beston 2001; MAF 2002a, 2002d, 2002e, 2002g). Moreover, the Ombudsman (2007) estimated 43,000 properties were affected during the height of the spraying in January 2003. In turn, as the number of affected properties gradually grew, so did the number of people being exposed to the spraying.

The number of sprayings also grew incrementally. Where MAF originally announced the spraying operation would consist of six to eight sprayings, in May they announced they would continue spraying during the New Zealand Winter. In August 2002 the re-elected Labour government approved ten additional large-scale aerial sprayings for 2002–2003 (Office of the Ombudsman 2007). Moreover, after completing those additional sprayings, in May 2003 they decided to add an 11th spraying prior to the Winter months (i.e. June to August) and to resume spraying in Spring, without specifying the full number of sprayings that would be administered (*Ibid.*). By the end of the eradication operation, MAF conducted 40 aerial sprayings, which consisted of applying pesticides on 60 different days (Goven et al. 2007).

The operation's cumulative expansion served as another powerful controlling process. Although it is hard to imagine that aerial pesticide spraying over cities would ever be considered "normal," the incremental process through which New Zealand officials expanded the operation enabled the expansion to pass more undetected than would have been the case if government officials had announced the full scope of the operation at the outset. Moreover, expanding the operation in a less detectable way minimized the number of new people who would be alarmed by each expansion and limited the number of new people who would become inclined to criticize and oppose the operation and/or government agencies running it. In turn, minimizing the potential for opposition preserved the operation's institutional momentum.

An implication of this case is that when analyzing the rollout of urban pesticide spraying operations, it is important to note the government's use of cumulative tinkering, identifying which aspects of the operation they tinkered with and how they benefit from its use.

Portraying the Pesticide as Harmless to Humans

In her analysis of chemical production communications, Sarah Vogel (2012) emphasizes that product safety perceptions are highly malleable and should be seen as social constructions that need to be analyzed and unpacked. Moreover, scholars have shown that much of that social construction is done by product manufacturers (such as tobacco and chemical manufacturers), who deploy numerous strategies and tactics to develop and reinforce the perception their products are safe (Déplaude 2015; Markowitz & Rosner 2002; Proctor 2008; Vogel 2012).

Beyond manufacturers, we also need to consider the role government agencies play in shaping product safety perceptions, as they have the resources to produce sophisticated communication campaigns, enjoy ample access to mass media, and the public tends to grant them the power to speak authoritatively on such matters.

As this pertains to the PAM spraying operation, my analysis of the New Zealand government's communications campaign[1] revealed it portrayed the 48B pesticide as being safe by making three different but related claims: (1) Foray 48B is harmless to humans; (2) its main bacterial ingredient (Btk) is harmless to humans; and (3) the added synthetic chemicals are harmless to humans.

In this section I document how government agents advanced these claims and highlight problems with those claims, which includes the fact they glossed over evidence that ran counter to them. I also discuss how, collectively, the deployment of these claims helped allay safety concerns about the spraying operation, thereby also slowing the spread of opposition to the pesticide spraying.

Portraying the Foray 48B as Harmless to Humans

In the weeks leading up to the first spraying in mid-January (MAF 2001e) government officials made numerous public statements that suggested Foray 48B was harmless to humans. First, on December 11, 2001, the Auckland District Health Board (ADHB) declared in the *New Zealand Herald* (Auckland's largest newspaper) that "Aerial spraying to eradicate the imported painted apple moth pest will not be hazardous to the health of residents" (New Zealand Herald 2001). This statement was reinforced the following week by a MAF (2001f) media release that made the following two statements:

> The Auckland District Health Board recently released a draft Health Risk Assessment for consultation. The assessment gave the Foray 48B spray to be used in the moth eradication programme the all clear. It said it did not expect any significant health effects from the spray. After 35 years of use Foray 48B has never been implicated in human infection.

On December 21st the messaging was further reinforced when ADHB publicly disseminated a fact sheet with the bold title: "No significant health effects are expected from the spray programme" (Kelly 2001). The following month, on the first day of spraying, a MAF (2002a) press release communicated the following statements:

> An independent health risk assessment carried out by the Auckland District Health Board concluded that Foray 48B has never been implicated in human infection or any other significant health condition. No special health precautions need to be taken.

Further reinforcing the message was that week's ADHB media release, which claimed the spray has a "clean bill of health" and "is harmless to humans and animals" (Kelly 2002).

Similar messaging occurred throughout the spraying operation. In July 2002 a media release from the Minister of Biosecurity (Jim Sutton) extolled Foray 48B's "proven safety record" (New Zealand Government 2002). In November, when MAF was expanding the spray zone by 9% (from 7,980 to 8,686 hectares) the PAM Operation's general manager appeared on television to reiterate the safety messaging (Goven et al. 2007; Isbister 2002). In March 2003, when MAF was facing increased pressure resulting from the Blackmore (2003) report, the agency bolstered the safety messaging by placing full page ads in *The New Zealand Herald* that contained the following headline: "People are asking is there any evidence of long-term health effects resulting from the spray? The answer is no" (2003). As well, in July of that year, after the government decided to unexpectedly extend the spraying campaign a second time,

MAF reiterated the safety message by disseminating a "fact" sheet featuring the same title as the March advertisement (Office of the Ombudsman 2007: 59).

Unfortunately, there is significant evidence suggesting the government's glowing portrayal of Foray 48B was incomplete and misleading. First, Foray 48B was known to cause allergic reactions and sensitization, as underscored in the manufacturer's 1991 Material Safety Data Sheet, which states "Repeated exposure via inhalation can result in sensitization and allergic response in hypersensitive individuals" (Novo Nordisk 1991, as cited in Swadener 1994). This should have been an important consideration for government officials right from the start, particularly when we consider that 13,500 residents were in the initial spray zone and that this area was going to be sprayed between six and eight times (Blackmore 2020; Office of the Ombudsman 2007).

Second, there was documented evidence that previous uses of Foray 48B, both abroad and in Auckland, were associated with a range of health problems, including respiratory, digestive, and neurological ailments. For instance, ground spray applicators in Vancouver (Canada) reported eye, nose, throat, and respiratory irritation (Nobel et al. 1992). Health complaints also emerged in urban areas where Foray 48B was aerially sprayed. According to the Washington state health department 250 people reported health problems following the 1993 spraying over Spokane, with another 59 reporting health problems following the 2000 spraying operation over Seattle (Washington State Department of Health 1993, 2001). Health complaints were also reported over the course of the 1996–1997 spraying operation against the White-Spotted Tussock Moth (which also used Foray 48B) in East Auckland, where a community health monitoring program identified 375 residents who suffered health problems attributed to the spraying, including respiratory symptoms (such as asthma, chest tightness, and coughs), headaches, skin irritation, skin rashes, sore throats, blocked noses, eye irritation, diarrhea, vomiting, stomach cramps, flu-like symptoms (such as fever, malaise, and swollen glands), and lethargy (Hales et al. 2004; Office of the Ombudsman 2007; Watts 2003). Additionally, a Ministry of Forestry study revealed that 8% of East Auckland residents reported being affected by that spraying program, and that the figure went up to 9.9% in the more frequently sprayed areas and 16% in the most sprayed areas (Allpro Consulting of Wellington 1997; Auckland Healthcare 1997; Blackmore 2020).

Further evidence emerged during and after the first year of spraying. First, in November and December 2002 a local pharmaceutical manufacturer administered surveys that revealed 15% of employees experienced spray effects or had family members who did. Second, at the end of 2002 a community health monitoring program identified 315 residents who reported respiratory, digestive, and neurological problems during the spraying, all consistent with health issues that emerged during the 1996–1997 spraying operation in East Auckland (Blackmore 2003). It is noteworthy that the government had the resulting report (the Blackmore report) reviewed by an independent expert, who judged the report's methodology to be appropriate, the reasoning sound, and that the report was a credible source of knowledge (Phillips 2003; Office of the Ombudsman 2007). Moreover, in February 2003 the New Zealand Institute of Education surveyed staff working at schools in the spray zone, with 56% (197 of 353) of respondents reporting they were adversely affected by the spray *and* had noted similar effects among children (Office of the Ombudsman 2007). The latter evidence, in particular, should have suggested it was not just hypersensitive people that were being affected by the spray.

In addition, further evidence emerged *after* the PAM operation's conclusion, as 3,888 householders were reported to have gone to the Auckland spray health service with multiple symptoms, including respiratory, skin, digestive, eye, and neurological problems (Office of the Ombudsman 2007). Additionally, there was a statistically significant increase in monthly asthma discharge rates for children living in the spray zone: a 100% increase for boys aged 0–4 and an 80% increase for girls aged 5 and 14 (Gallagher et al. 2005).

The government's rosy portrayal of the pesticide was not lost on the Ombudsman, who argued "MAF made a big error early on in the spray program by giving the impression that the spray had no health effects" (2007: 9–10). Additionally, he recommended that government agencies who carry out spraying operations in the future should "unequivocally acknowledge that there may be harm caused to some people residing or present within the spray zone" (*Ibid.*).

Despite significant evidence to the contrary, MAF and health officials maintained Foray 48B was harmless to humans, a narrative they disseminated before, during and even after the conclusion of the spraying operation. In turn, maintaining this narrative also allayed citizen concerns and helped reduce the potential for citizens to turn against the spraying operation. This was particularly true for those outside the spray zones, who had no first-hand knowledge of the spraying.

Portraying Btk as Harmless to Humans

Beyond portraying Foray 48B as harmless to humans, government officials also claimed the pesticide's main ingredient (i.e. the *Bacillus thuringiensis subspecies kurstaki* (Btk) bacteria) was harmless. They did this by making three sorts of claims: (1) Btk only harms caterpillars; (2) the bacteria cannot survive or become active in warm-blooded animals; and (3) there is no evidence it has harmed humans.

Claim: Btk Only Targets Caterpillars

The first claim is exemplified by the following statement, which appeared in an October 2001 MAF press release on the same day they announced they would be escalating PAM eradication efforts by pursuing aerial pesticide spraying over parts of Auckland: "Btk, when applied at recommended rates, does not harm people, plants, animals or any insects – except for caterpillars" (MAF 2001a).

While this statement was geared towards allaying citizen concerns, it was at odds with the existing evidence, which showed that Btk was capable of harming many other species. For example, in other contexts the use of Bt-based sprays decreased numerous beneficial insects, including parasitic wasps, spider mites, and aphid-eating flies (Swadener 1994). Such sprays have also been shown to be directly toxic to water insects (Kreutzueiser et al. 1992), Rainbow trout (Swadener 1994), and some birds (Jones 1986).

Land mammals have also been affected. In one study rats experienced respiratory depression when exposed to air containing Btk spores (Swadener 1994) and in another rabbits exposed to Bt experienced irritation to the skin and eyes (Seigel et al. 1987). As well, sheep who were exposed to Btk through diet had loose stools, with some displaying microscopic damage to their colons (Hadley et al. 1987). Moreover, during the PAM spraying campaign there were over 20 reports of cats being unwell following sprayings, with typical symptoms including vomiting, lack of hunger, infected eyes, and skin allergies (Blackmore 2003). There were also reports of dogs being affected, with the most common symptom being diarrhea, followed by eye and skin problems (*Ibid.*). A potential reason for these effects is that while Bt is indeed naturally found in the environment, Btk is a relatively rare strain, and so many species might be unaccustomed to handling it. Moreover, the strains used in the Foray 48B pesticide have been engineered, sometimes genetically so, to be up to six times more potent than would naturally be found in the environment (No Spray Zone n.d.; Sanahuga et al. 2011).

Claim: The Bacteria Can't Survive in Warm-Blooded Animals

Beyond suggesting Btk only targets caterpillars, MAF claimed this was the case because the bacteria could not survive in warm-blooded animals, as exemplified by the following quote:

> Btk has been chosen as a preferred spray for aerial application against painted apple moth because it specifically targets caterpillars, does not grow in warm-blooded organisms (animals or humans)...
>
> (MAF 2001a)

This statement is also problematic as the aforementioned data on animals suggests the bacteria can survive in warm-blooded animals. Additionally, there is research suggesting it is also true for humans. Specifically, researchers have found that Bt spores can survive and remain viable in humans for months after exposure (Bernstein et al. 1999; Valadares et al. 2001). As well, subsequent research found that 25% of exposed humans manifest immune responses that can last up to three years, which suggests the bodies are fighting off a reproducing population (Doekes et al. 2004). Moreover, while these issues are particularly concerning for those with impaired immune systems,

Table 7.1 Chemicals Used in Various Foray 48B Formulations

Compound	CAS Number	Potential Health Problems Associated with the Compound
1,5-Hexanediene-3,4-diol,2,5-dimethyl	4723-10-8	
1-Propanesulfonyl chloride	10147-36-1	Inhalation of this product "is extremely destructive to the tissue of the mucous membranes and upper respiratory tract. Symptoms may include coughing and shortness of breath, and it may cause headaches and nausea" (European Chemicals Agency 2020a). Additionally, ingestion "may cause burns in mouth and throat," skin contact "causes severe skin burns," and eye contact is corrosive, causing stinging, tearing, and redness (*Ibid.*).
4-Acetyloxy-2-butanone	10150-87-5	Causes skin irritation, causes serious eye irritation, and may cause respiratory irritation (European Chemicals Agency 2020b)
2-Methyl-2,3-pentanediol	7795-80-44	
2,4-Hexadienedioic acid	505-70-4	Suspected to be hazardous to aquatic life and "suspected skin sensitiser" (European Chemicals Agency 2020c)
2-Heptanone, 3-hydroxy-3-methyl	13757-91-0	
2-Hydroxy pyridine	142-08-5	"This substance is toxic if swallowed, causes serious eye irritation, causes skin irritation and may cause respiratory irritation" (European Chemicals Agency 2020d).
5-Hexen-2-one, 5-methyl	3240-09-3	"This substance is a flammable liquid and vapour, is harmful if swallowed and causes serious eye damage" (European Chemicals Agency 2020e).
Acetic acid, 2-propenyl ester	591-87-7	Glacial acetic acid is toxic to humans and animals by inhalation and skin contact. Humans exposed to 1000 ppm for a few minutes may suffer eye and respiratory tract irritation (Patnaik 2007).
Acetic acid, anhydride	108-24-7	"This substance causes severe skin burns and eye damage, is a flammable liquid and vapour, is harmful if swallowed and is harmful if inhaled" (European Chemicals Agency 2020f)
Benzoic acid[a]	65-85-0	Causes asthma and irritates the skin and eyes (Brunekeef & Holgate 2002; Vizcaya et al. 2011). Toxic symptoms include somnolence, respiratory depression, and gastrointestinal disorder (Patnaik 2007). "This substance causes damage to organs through prolonged or repeated exposure, causes serious eye damage and causes skin irritation" (European Chemicals Agency 2020g)
Butylated hydroxy toluene	128-37-0	"This substance is very toxic to aquatic life with long lasting effects" and is under assessment for endocrine disruption (European Chemicals Agency 2020h)
Cyclohexasiloxane, dodecamethyl	540-76-6	
Decamethyl cyclopentasiloxane	541-02-6	This substance is "persistent, bio-accumulative and toxic" (European Chemicals Agency 2020i)

Compound	CAS Number	Potential Health Problems Associated with the Compound
Cyclotetrasiloxane, octamethyl	556-67-2	This substance is "persistent, bio-accumulative and toxic." Additionally, it "is a flammable liquid and vapour, is suspected of damaging fertility or the unborn child and may cause long lasting harmful effects to aquatic life" (European Chemicals Agency 2020j)
Cyclotrisiloxane, hexamethyl	541-05-9	This "substance causes serious eye irritation, may cause respiratory irritation and causes skin irritation" (European Chemicals Agency 2020k)
Disiloxane derivative	18420-09-2	"This substance is a flammable liquid and vapour, causes serious eye irritation, causes skin irritation and may cause respiratory irritation" (European Chemicals Agency 2020l)
Hydrochloric acid[a]	7647-01-0	"This substance causes severe skin burns and eye damage, is toxic if inhaled, may damage fertility or the unborn child, causes serious eye damage, may cause damage to organs through prolonged or repeated exposure, may be corrosive to metals and may cause respiratory irritation" (European Chemicals Agency 2020m). Additionally, inhalation can cause choking and inflammation of the respiratory tract, and the substance is linked to asthma, lung cancer, and bronchitis (Brunekeef & Holgate 2002; CHE 2019a; Patnaik 2007; Vizcaya et al. 2011).
Methyl paraben	99-76-3	This substance is toxic to aquatic life and is being assessed for endocrine disruption (European Chemicals Agency 2020n). Also, it is linked to: (1) Skin irritation and allergy; and (2) increases in breast cancer tumor proliferation (Ishiwatari et al. 2006; Lillo et al. 2016; Matwiejczuk et al. 2020).
Penta siloxane, dodecamethyl	141-63-9	
Phenyl amine R silane derivative	10538-85-9	
Trimethyl phosphine	594-09-2	"This substance is a highly flammable liquid and vapour, causes serious eye irritation, causes skin irritation and may cause respiratory irritation" (European Chemicals Agency 2020o). Also, preliminary evidence links phosphine to cataracts, chronic renal disease, heart attacks, and peripheral neuropathy (CHE 2019b)
Phophoric acid	7664-38-2	(1) Dermatological exposure can irritate skin and mucous membranes; (2) its vapors can cause coughing and throat irritation (Patnaik 2007)
Potassium phosphate	7778-53-2	"This substance causes serious eye damage and may cause respiratory irritation (European Chemicals Agency 2020p)
Propylene glycol[a]	57-55-6	Linked to hearing loss, skin irritation, intestinal damage, depression, and has been found to affect children's central nervous systems (CHE 2019c; Kedgley 2003)

Sodium hydroxide	1310-73-2	"This substance causes severe skin burns and eye damage" (European Chemicals Agency 2020q). Additionally, it is: (1) Severely corrosive to eyes, skin, mucous membranes and digestive systems; (2) breathing sodium hydroxide dust or mist leads in mild cases to irritation of the mucous membranes of the nose... and in severe cases to damage of the upper respiratory tract" (Harte et al. 1991)
Sulfuric acid	7664-93-9	(1) Can cause severe skin burns and permanent vision loss; (2) inhaling it as a mist or vapors can produce coughing and significant bronchial constriction; and (3) chronic exposure can produce bronchitis, conjunctivitis, skin lesions, and erosion of teeth; (4) cause laryngeal cancer (CHE 2019d; Patnaik 2007)
Sydnone, 3-phenylmethyl	16844-42-1	
Thietane	287-27-4	"This substance is a highly flammable liquid and vapour and is harmful if swallowed" (European Chemicals Agency 2020r)
Trisiloxane	3555-47-3	

[a] Chemical was confirmed to be in the formulation used in the PAM operation.

This arrangement served as another powerful controlling process as it made the public *almost*[2] completely reliant on government officials for knowledge about which ingredients were in the pesticide formulation and what health problems were associated with each ingredient. This made it almost impossible for the public to adequately evaluate the government's claims that the ingredients were harmless. The best they could do was investigate the chemicals used in previous Foray 48B formulations (see Table 7.1 for a list of some of those chemicals), identify the health problems associated with those chemicals, and hope no new chemicals were added to the formulation they were being exposed to. In turn, the situation also helped slow the spread of opposition to the pesticide, the spraying operation, or the agency running the operation.

As a whole, the government communication campaign conveyed the impression Foray 48B was harmless to humans and that no special precautions needed to be taken. Moreover, there is evidence to suggest the campaign's messaging was effective: a government survey conducted in April 2003 (16 months into the operation and the month before the government extended the spraying operation a second time) revealed that 79% of randomly sampled adults could recall ads conveying that few people would be affected by the spraying (Office of the Ombudsman 2007: 50). Consequently, future research on urban aerial pesticide spraying operations should pay careful attention to the communication campaigns that government agencies deploy, scrutinizing the safety claims they make about both the pesticide and its ingredients, as well as analyzing the degree to which the communications address evidence about harms caused by the whole pesticide or any of its ingredients.

Summary

This chapter illuminated three government activities that reduced citizen concerns about the PAM pesticide spraying operation: (1) proposing a spraying operation that was of far smaller scope than what it would eventually expand to; (2) expanding the spraying operation incrementally; and (3) portraying the pesticide as harmless to humans. In doing so, this chapter highlights some of the government actions that can allay concerns about pesticide use, and, in turn, slow the spread of opposition to such spraying operations. Beyond helping us better understand what can slow opposition to spraying operations to eradicate foreign species, the analysis can also be useful for understanding the controlling processes that hinder opposition to other forms of urban pesticide use, including spraying on playgrounds, parks, sidewalks, sports fields, golf courses, and roads.

Having said this, there is more to the story. Although portraying a pesticide as harmless is a powerful government tactic to allay pesticide concerns, it is important to note its effectiveness will be mediated by the presence or absence of knowledge that could undermine the portrayal, such as evidence that the pesticide can harm people. Working to suppress or downplay such evidence can substantially enhance government claims that pesticides are harmless, and elucidating this important issue is what we now turn to.

Notes

1 The analysis reviewed all government communications related to the pesticide used in the PAM aerial spraying operation, which included all press releases disseminated by government agents, including 104 released by the Ministry of Agriculture and Forestry. The analysis also included government reports as well as statements that government agents communicated via mass media.

2 I say "almost" because researchers at the University of British Columbia became an alternative source of knowledge about the Foray 48B pesticide, when they reversed engineered a previous formulation to reveal a handful of the chemicals that were used. Another source of alternative information was the New Zealand Green Party, which obtained a partial list of the ingredients from activists and tabled them (Green Party 2003).

References

Allpro Consulting of Wellington. 1997. *Tussock Moth Spraying Programme Survey 111.* July 11, 1997.

Auckland District Health Board, Public Health Service. 2002. *Health Risk Assessment of the 2002 Aerial Spray Eradication Programme for the Painted Apple Moth in some Western Suburbs of Auckland* (p. 8).

Auckland Healthcare. 1997. *Health Risk Assessment of the proposed 1996–1997 Control Programme for the White-Spotted Tussock Moth in the Eastern Suburbs of Auckland.* A Report to the Ministry of Forestry by the Public Health Protection Service, Auckland Healthcare Services Ltd. September.

Beston, Anne. 2001. "Big Spray targets Western suburbs." *New Zealand Herald*, November 9. https://www.nzherald.co.nz/nz/big-spray-targets-western-suburbs/YJWBY2UBNKXHRZ26HFBKWQHNFY/

Bernstein, Leonard, Jonathan Bernstein, Maureen Miller, Sylva Tierzieva, David Bernstein, Zana Lummus, MaryJane Selgrade, Donald Doerfler, and Verner Seligy. 1999. "Immune responses in farmer workers after exposure to Bacillus thuriengiensis pesticides." *Environmental Health Perspectives* 107(7): 575–582.

Blackmore, Hana. 2003. *Interim Report of the Community-based Health and Incident Monitoring of the Aerial Spray Programme*. Auckland, New Zealand.

Blackmore, Hana. 2020. *Btk Pesticide Sprays: Adverse Health Impacts and Effects: Cover up or Abdication of Responsibility?* Unpublished report. Auckland, New Zealand.

Brunekeef, Bert and Stephen Holgate. 2002. "Air Pollution and Health." *Lancet* 360(9341): 1233–1242.

CHE (Collaborative on Health and the Environment). 2019a. *Toxicant and Disease Database Entry for 'Hydrochloric Acid'*. Accessed February 1, 2022. https://www.healthandenvironment.org/our-work/toxicant-and-disease-database/?showcategory=&showdisease=&showcontaminant=2651&showcas=&showkeyword=

CHE (The Collaborative on Health and the Environment). 2019b. *Toxicant and Disease Database Entry for 'Phosphine'*. Accessed February 1, 2022. https://www.healthandenvironment.org/our-work/toxicant-and-disease-database/?showcategory=&showdisease=&showcontaminant=2364&showcas=&showkeyword=

CHE (Collaborative on Health and the Environment). 2019c. *Toxicant and Disease Database Entry for 'Propylene Glycol'*. February 1, 2022. https://www.healthandenvironment.org/our-work/toxicant-and-disease-database/?showcategory=&showdisease=&showcontaminant=2995&showcas=&showkeyword=

CHE (Collaborative on Health and the Environment). 2019d. *Toxicant and Disease Database Entry for 'Sulfuric Acid'*. February 1, 2022. https://www.healthandenvironment.org/our-work/toxicant-and-disease-database/?showcategory=&showdisease=&showcontaminant=2798&showcas=&showkeyword=

Coppin, Dawn M., Brian W. Eisenhauer, and Richard S. Krannich. 2002. "Is Pesticide Use Socially Acceptable? A Comparison between Urban and Rural Settings." *Social Science Quarterly* 83(1): 379–393.

Damgaard, Per Hyldebrink. 1995. "Diarrhoeal Enterotoxin Production by Strains of Bacillus thuringiensis Isolated from Commercial Bacillus thuringiensis-based Pesticides." *FEMS Immunology and Medical Microbiology* 12: 245–250.

Déplaude, Marc-Olivier. 2015. "Minimising Dietary Risk: The French Association of Salt Producers and the Manufacturing of Ignorance." *Health Risk & Society* 17(2): 168–183.

Doekes, Gert, Preben Larsen, Torben Sigsgaard, and Jesper Baelum. 2004. "IgE Sensitization to Bacterial and Fungal Biopesticides in a Cohort of Danish Greenhouse Workers: The BIOGART Study." *American Journal of Industrial Medicine* 46(4): 404–407.

Dunlap, Riley E. and Curtis E. Beus. 1992. "Understanding Public Concerns about Pesticides: An Empirical Examination." *Journal of Consumer Affairs* 26(2): 418–438.

European Chemicals Agency. 2020a. *Entry for "1-Propanesulfonyl Chloride" in the REACH Database*. https://echa.europa.eu/registration-dossier/-/registered-dossier/30670/9

European Chemicals Agency. 2020b. *Entry for "4-Acetyloxy-2-Butanone" in the REACH Database*. https://echa.europa.eu/information-on-chemicals/cl-inventory-database/-/discli/notification-details/154301/608336

European Chemicals Agency. 2020c. *Entry for "2,4-Hexadienedioic Acid" in the REACH Database.* https://echa.europa.eu/information-on-chemicals/annex-iii-inventory/-/dislist/details/AIII-100.007.289

European Chemicals Agency. 2020d. *Entry for "2-Hydroxy Pyridine" in the REACH Database.* https://echa.europa.eu/substance-information/-/substanceinfo/100.005.019

European Chemicals Agency. 2020e. *Entry for "5-Methylhex-5-en-2-One" in the REACH Database.* https://echa.europa.eu/substance-information/-/substanceinfo/100.019.825

European Chemicals Agency. 2020f. *Entry for "Acetic Acid, Anhydride" in the REACH Database.* https://echa.europa.eu/substance-information/-/substanceinfo/100.003.241

European Chemicals Agency. 2020g. *Entry for "Benzoic Acid" in the REACH Database.* https://echa.europa.eu/substance-information/-/substanceinfo/100.000.562

European Chemicals Agency. 2020h. *Entry for "Butylated Hydroxy Toluene" in the REACH Database.* https://echa.europa.eu/substance-information/-/substanceinfo/100.004.439

European Chemicals Agency. 2020i. *Entry for "Decamethylcyclopentasiloxane" in the REACH Database.* https://echa.europa.eu/substance-information/-/substanceinfo/100.007.969

European Chemicals Agency. 2020j. *Entry for "OctamethylCyclotetrasiloxane" in the REACH Database.* https://echa.europa.eu/substance-information/-/substanceinfo/100.008.307

European Chemicals Agency. 2020k. *Entry for "HexamethylCyclotrisiloxane" in the REACH Database.* https://echa.europa.eu/brief-profile/-/briefprofile/100.007.970#collapseSeven

European Chemicals Agency. 2020l. *Entry for "Disiloxane Derivative" in the REACH Database.* https://echa.europa.eu/substance-information/-/substanceinfo/100.038.438

European Chemicals Agency. 2020m. *Entry for "Hydrochloric Acid" in the REACH Database.* https://echa.europa.eu/substance-information/-/substanceinfo/100.028.723

European Chemicals Agency. 2020n. *Entry for "Methyl Paraben" in the REACH Database.* https://echa.europa.eu/substance-information/-/substanceinfo/100.002.532

European Chemicals Agency. 2020o. *Entry for "TrimethylPhosphine" in the REACH Database.* https://echa.europa.eu/substance-information/-/substanceinfo/100.008.932

European Chemicals Agency. 2020p. *Entry for "Potassium Phosphate" in the REACH Database.* https://echa.europa.eu/substance-information/-/substanceinfo/100.029.006

European Chemicals Agency. 2020q. *Entry for "Sodium Hydroxide" in the REACH Database.* https://echa.europa.eu/substance-information/-/substanceinfo/100.013.805

European Chemicals Agency. 2020r. *Entry for "Thietane" in the REACH Database.* https://echa.europa.eu/substance-information/-/substanceinfo/100.005.469

Fisher Scientific. 2000. *Material Safety Data Sheet: Propylene Glycol.* http://www.fishersci.com

Gallagher, Lou, Ruth Pirie, and Simon Hales. 2005. *Descriptive Study of Hospital Discharges for Respiratory Diseases in Spray Zone for Painted Apple Moth (Auckland), Relative to Local and National Statistics 1999–2004.* Prepared as part of a Ministry of Health contract for scientific services.

Goven, Joanna, Tom Kerns, Romeo Quijano, and Dell Wihongi. 2007. *Report of the March 2006 People's Inquiry Into the Impacts and Effects of Aerial Spraying Pesticide over Urban Areas of Auckland.* Auckland, NZ: Action Plan & Print. https://peoplesinquiry.wordpress.com/

Green Party. 2002a. "Moth programme falling to pieces as head resigns." *Scoop Independent News,* May 09.

Green Party. 2002b. "Greens disappointed community moth plan ignored." *Scoop Independent News,* September 11. https://www.scoop.co.nz/stories/PA0209/S00176.htm

Green Party. 2003. "DIY detectives reveal 'secret' spray ingredients." *Scoop Independent News,* May 5. https://www.scoop.co.nz/stories/PA0305/S00087.htm

Hadley William M., Scott W. Burchiel, Thomas D. McDowell, John P. Thilsted, Clair M. Hibbs, Jerry A. Whorton, Phillip W. Day, Mitchell B. Friedman, and Raymond E. Stoll. 1987. "Five-Month Oral (Diet) Toxicity/Infectivity Study of *Bacillus thuringiensis* Insecticides in Sheep." *Fundamental and Applied Toxicology* 8(2): 236–242.

Hales, Simon, Virginia Baker, Kevin Dew, Losa Moata'ane, Jennifer Martin, Tim Rochford, David Slaney, and Alistair Woodward. 2004. *Assessment of the Potential Health Impacts of the 'Painted Apple Moth' Aerial Spraying Programme, Auckland.* Report for the New Zealand Ministry of Health.

Harte, John, Cheryl Holden, Richard Schneider, and Christine Shirley. 1991. *Toxics A to Z: A Guide to Everyday Pollution Hazards.* Berkeley: University of California Press.

INCHEM. 2000. *Benzoic Acid and Sodium Benzoate.* International Programme on Chemical Safety. http://www.inchem.org/documents/cicads/cicads/cicad26.htm.

Isbister, Robert. 2002. *Holmes Show.* TV One, November 20.

Ishiwatari, S., T. Suzuki, T. Hitomi, T. Yoshino, S. Matsukuma, and T. Tsuji. 2007 "Effects of Methyl Paraben on Skin Keratinocytes." *Journal of Applied Toxicology* 27(1): 1–9.

Jones, I. W. 1986. *Summary Report: Effect of Dipel and Plyac on Hatchability of Ringneck Pheasant Eggs.* Oregon Dept. of Fish and Wildlife.

Kedgley, Sue. 2003. "Toxic ingredients of Painted Apple Moth spray tabled." *Green Party Media Release,* May 7. http://www.greens.org.nz/speeches/toxic-ingredients-painted-apple-moth-spray-tabled

Kelly, Francesca. 2001. *Foray 48B Spraying in West Auckland (The Painted Apple Moth Eradication Programme).* Public Health Protection Office, Auckland District Health Board. December 21.

Kelly, Francesca. 2002. "Painted apple moth project health team up and running." *ADHB Media release,* January 14.

Kreutzueiser, David P., Stephen B. Holmes, Scott S. Capell, and David C. Eichenberg. 1992. "Lethal and Sublethal Effects of Bacillus thuringiensis var. kurstaki on Aquatic Insects in Laboratory Bioassays and Outdoor Stream Channels." *Bulletin of Environmental Contamination and Toxicology* 49: 252–258.

Lillo M. Angeles, Cydney Nichols, Chanel Perry, Stephanie Runke, Raisa Krutilina, Tiffany N. Seagroves, Gustavo A. Miranda-Carboni, and Susan Krum. 2016. "Methylparaben stimulates tumor initiating cells in ER+ breast cancer models." *Journal of Applied Toxicology* 37(4): 417–425.

Markowitz, Gerald and David Rosner. 2002. *Deceit and Denial: The Deadly Politics of Industrial Pollution.* Berkeley: University of California Press.

Matwiejczuk, Natalia, Anna Galicka, and Malgorzata Brzózka. 2020. "Review of the Safety of Application of Cosmetic Products Containing Parabens." *Journal of Applied Toxicology* 40(1): 176–210.

Ministry of Agriculture and Forestry (MAF). 2000. *Potential Economic Impact on New Zealand of the Painted Apple Moth.* Wellington, NZ.

Ministry of Agriculture and Forestry (MAF). 2001a. "About the Btk Spray." *Scoop–Independent News*, October 25.

Ministry of Agriculture and Forestry (MAF). 2001b. "Targeted aerial spraying to go ahead in West Auckland." *Scoop–Independent News*, October 25.

Ministry of Agriculture and Forestry (MAF). 2001c. "No evidence of health problems from Btk use." *Scoop–Independent News*, October 25.

Ministry of Agriculture and Forestry (MAF). 2001d. "Spray programme hits snag – Aerial spray setback." *Scoop–Independent News*, November 27.

Ministry of Agriculture and Forestry (MAF). 2001e. "New plans for target aerial spraying." *Scoop–Independent News*, December 13.

Ministry of Agriculture and Forestry (MAF). 2001f. "More medical help for west Aucklanders." *Scoop–Independent News*, December 17.

Ministry of Agriculture and Forestry (MAF). 2002a. "Spray one against moth completed." *Scoop–Independent News*, January 24.

Ministry of Agriculture and Forestry (MAF). 2002b. "Faster aerial spray operation planned." *Scoop–Independent News*, May 6.

Ministry of Agriculture and Forestry (MAF). 2002c. "High winds and rain stop spraying again." *Scoop–Independent News*, August 18.

Ministry of Agriculture and Forestry (MAF). 2002d. "Ninth spray date against moth pest set." *Scoop–Independent News*, September 4.

Ministry of Agriculture and Forestry (MAF). 2002e. "Painted Apple Moth Eradication Project: Fact Sheet." *Scoop–Independent News*, September 10.

Ministry of Agriculture and Forestry (MAF). 2002f. "Painted apple moth meeting." *Scoop–Independent News*, October 1.

Ministry of Agriculture and Forestry (MAF). 2002g. "Still no go for aerial operation against moth." *Scoop–Independent News*, October 10.

Ministry of Agriculture and Forestry (MAF). 2002h. "Map: Painted Apple Moth Aerial Zone Extended." *Scoop–Independent News*, November 21.

Ministry of Agriculture and Forestry (MAF). 2003. "People are asking is there any evidence of long-term health effects resulting from the spray?" *New Zealand Herald*, March 11.

Nader, Laura. 1997. "Controlling Processes: Tracing the Dynamic Components of Power." *Current Anthropology* 38(5): 711–738.

New Zealand Government. 2002. "Painted Apple Moth." *Scoop–Independent News*, July 3. https://www.scoop.co.nz/stories/PA0207/S00117.htm

New Zealand Herald. 2001. "Painted apple moth spray will not affect health, says DHB." *New Zealand Herald*, December 11.

New Zealand Herald. 2003. "Moth spraying will threaten growers' organic status." *New Zealand Herald*, August 13. https://www.nzherald.co.nz/nz/news/article.cfm?c_id=1&objectid=3517901

New Zealand Labour Party. 2001. "No blanket spraying." *Scoop Independent News*, November 12. https://www.scoop.co.nz/stories/PA0111/S00214.htm

No Spray Zone. n.d. BT: Organic pesticide or environmental disaster. Accessed January 30, 2016. http://nosprayzone.number6.org/bt-info/bt-organic-pesticide-or-environmental-disaster/

Noble, Michael A., Peter D. Riben, and Gregory J. Cook. 1992. *Microbiological and Epidemiological Surveillance Program to Monitor the Health Effects of Foray 48B BTK Spray*. Vancouver, BC: Ministry of Forests. Province of British Columbia.

Novo Nordisk. 1991. *Material Safety Data Sheet for Foray 48B Flowable Concentrate.* Danbury, CT.

Office of the Controller and Auditor-General. 2002. *Report of the Controller and Auditor-General: Management of Biosecurity Risks: Case Studies.* Wellington, NZ: The Audit Office. https://oag.parliament.nz/2002/biosecurity-case-studies/docs/part3.pdf

Office of the Ombudsman (Mel Smith). 2007. *Report of the Opinion of Ombudsman Mel Smith on Complaints Arising from Aerial Spraying of the Biological Insecticide Foray 48B on the Population of Parts of Auckland and Hamilton to Destroy Incursions of Painted Apple Moths, and Asian Gypsy Moths, Respectively.* Wellington, NZ: Office of the Ombudsman.

Painted Apple Moth Community Advisory Group (PAM CAG). 2001. "Get Ruth-Less with the Painted Apple Moth." *Scoop Independent News,* December 17. https://www.scoop.co.nz/stories/PO0112/S00077.htm

Panckhurst, Paul. 2001. "Mothbeaten." *Metro Magazine,* October: 57–63.

Patnaik, Pradyot. 2007. *A Comprehensive Guide to the Hazardous Properties of Chemical Substances.* Hoboken, NJ: John Wiley & Sons.

Phillips, David. 2003. "Commentary to the MoH on report titled: 'Draft Interim report of the Community-based health & incident monitoring of the aerial spray programme January – December 2002.'"

Proctor, Robert. 2008. "Agnotology: A Missing Term to Describe the Cultural Production of Ignorance (and Its Study)." In *Agnotology: The Making and Unmaking of Ignorance,* edited by Robert Proctor and Linda Schiebinger, 1–36. Stanford, CA: Stanford University Press.

Sanahuga, Georgina, Raviraj Banakar, Richard M. Twyman, Teresa Capell, and Paul Christou. 2011. *"Bacillus thuringiensis*: A Century of Research, Development and Commercial Applications." *Plant Biotechnology Journal* 9(3): 283–300.

Siegel, Joel P., John A. Shadduck, and James Szabo. 1987. "Safety of the Entomo-pathogen Bacillus thuringiensis var. Israeliensis for Mammals." *Journal of Economic Entomology* 80: 717–723.

Sutton, Jim. 2002. "Painted Apple Moth." Minister of Biosecurity media statement, July 3.

Swadener Carrie. 1994. *"Bacillus thuringiensis* (B.T.)." *Journal of Pesticide Reform* 14(3): 13–20.

Tayabali, Azam and Verner Seligy. 2000. "Human Cell Exposure Assays of Bacillus thuringiensis Commercial Insecticides: Production of Bacillus cereus-Like Cytolytic Effects from Outgrowth of Spores." *Environmental Health Perspectives* 108(10): 919–930.

Teschke, Kay, Yat Chow, Karen Bartlett, Andrew Ross, and Chris van Netten. 2001. "Spatial and Temporal Distribution of Airborne *Bacillus thuringiensis* Var. *kurstaki* during an Aerial Spray Program for Gypsy Moth Eradication. University of British Columbia, Canada." *Environmental Health Perspectives* 109(1): 47–54.

Valadares de Amorim, Giovana, Beatrixe Whittome, Benjamin Shore, and David Levin. 2001. "Identification of Bacillus thuriengiensis subsp. kurstaki Strain HD1-Like Bacteria from Environmental and Human Sample after Aerial Spraying of Victoria, BC, Canada with Foray 48B." *Applied and Environmental Microbiology* 67(3): 1035–1043.

Valent Biosciences. 2009. *Biological Insecticide Foray 48B Label.*

Vizcaya, David, Maria Mirabelli, Josep-Maria Antó, Ramon Orriols, Felip Burgos, Lourdes Arjona, and Jan-Paul Zock. 2011. "A Workforce-Based Study of Occupational Exposures and Asthma Symptoms in Cleaning Works." *Occupational and Environmental Medicine* 68(12): 914–919.

Vogel, Sarah. 2012. *Is It Safe? BPA and the Struggle to Define the Safety of Chemicals.* Berkeley: University of California Press.

Walsh, Frances. 2003. "Wipe Out." *Metro* March: 44–52.

Washington State Department of Health. 1993. *Report of Health Surveillance Activities: Asian Gypsy Moth Control Program.* Olympia, WA (March).

Washington State Department of Health. 2001. *Report of health surveillance activities: aerial spraying for Asian Gypsy Moth – May 2000 Seattle, WA.* https://www.doh.wa.gov/Portals/1/Documents/Pubs/334-292.pdf

Watts, Meriel. 2003. *Painted Apple Moth Eradication Programme: Health Risk and Effects.* https://www.semanticscholar.org/paper/PAINTED-APPLE-MOTH-ERAD-ICATION-PROGRAMME-%3A-HEALTH-Watts/758c4f98b73 b1e6190d09facf8639763ee68a805

West Aucklanders Against Aerial Spraying (WASP). 2002. "Govt Puts Health of Forest Before West Aucklanders." *Scoop Independent News*, September 10.

8 Managing Uncomfortable Knowledge

While portraying pesticides as harmless is important for building public acceptance of pesticide spraying, the success of such portrayals is predicated on effectively managing information that might undermine those portrayals. To better understand how government agents manage such information I used a synthetic explanatory framework that draws on both the "amplification and attenuation of risk" and the "social production of ignorance" literatures.

The "amplification and attenuation of risk" literature examines social forces that seek to mediate risk perceptions of potentially harmful products and the tactics they employ. Although this literature tends to focus on social forces that seek to *amplify* risk perceptions (such as activist groups and mass media) (Henderson et al. 2014; Kasperson & Kasperson 1996), an exception is Marc-Olivier Déplaude's (2015) work, which illuminates how other social forces (such as industry and government agencies) can work to *attenuate* risk perceptions of potentially harmful products. His analysis focuses on French salt manufacturers, who used various tactics (dissimulation, denial, diversion, undermining opponents, and intimidating opponents) to reduce the risk perceptions associated with salt consumption. In elucidating this case Déplaude (2015) provides a useful general framework for understanding the attenuation of risk perceptions that takes place for other potentially harmful products, such as pesticides.

For additional depth I also draw on the burgeoning literature addressing the social production of ignorance, which makes several contributions. First, it elucidates that while ignorance can consist of the absence of knowledge about a topic, it can also consist of false knowledge, where people hold erroneous information about a topic and/or give disproportionate attention to marginal or industry-funded research, as has been the case with climate change deniers (Michaels 2008; Proctor 2008). Second, it emphasizes that ignorance is more than a knowledge gap to be filled or a set of incorrect ideas to be corrected, as ignorance can also serve as a resource for those in power, cultivated to serve strategic purposes (McGoey 2012; Oreskes & Conway 2010; Proctor 2008;

DOI: 10.4324/9780429426414-9

Rayner 2012). For this reason, McGoey (2012) argues social scientists should focus less on the politics of knowledge and more on the

> politics of ignorance, the mobilization of ambiguity, the denial of unsettling facts, the realization that knowing the least amount possible is often the most indispensable tool for managing risks and exonerating oneself from blame in the aftermath of catastrophic events.
>
> (3)

The deliberate cultivation of ignorance signals that some knowledge is inconvenient and unsettling, what Rayner (2012) refers to as "uncomfortable knowledge." The production of ignorance literature also illuminates how such knowledge can be obscured from the public view, through the deployment of undone science, suppression, and neutralization tactics.

As this pertains to the PAM spraying operation, New Zealand government officials attenuated risk perceptions by deploying numerous tactics aimed at perpetuating ignorance about the pesticide, including delaying the commissioning of research that might yield uncomfortable results, circumscribing the aims of such research, and delaying the production and release of knowledge that might be inconvenient. Additionally, they neutralized knowledge that was already in the public sphere by using an assortment of strategies, including suppression, omission, dismissal, denial, downplaying, and diversion. As well, they stymied groups that could have provided uncomfortable knowledge.

Attenuating Risk Perceptions by Undone Science

Undone science, as illuminated through the work of Frickel et al. (2010) and Hess (2007), refers to the failure to authorize, fund, and/or complete research that some stakeholders consider to be essential. Undone science is particularly important for industries producing harmful products, as the absence of knowledge about harms and/or risks removes an important obstacle to successfully portraying their products as safe, which enables them to more effectively market their products and resist regulations.

An example is Ritalin, a controversial psychostimulant given to children for attention deficit hyperactivity disorder, which United States physicians began prescribing to children in the 1960s and whose worldwide consumption has skyrocketed over the last two decades (Diller 1999; Vallée 2018). This growth took place despite an absence of research and knowledge about the medication's long-term health impacts for most of this period, with the first results of long-term safety research emerging in the Netherlands, in 2017 (Solleveld et al. 2017). The manufacturers' unwillingness to conduct long-term safety studies perpetuated non-knowledge about the medication's potential long-term health problems, which made it easier to market the medication.

This tactic was also present during the PAM spraying operation, as the New Zealand government manifested significant reluctance to pursuing research that could yield uncomfortable knowledge about the pesticide's impacts. An example was the government's failure to systematically assess the health effects of the Foray 48B pesticide, either in the earlier 1996–1997 spraying operation against the white tussock moth, which also took place in Auckland, or in the first year of the PAM operation (Office of the Ombudsman 2007). Another example would be the government's unwillingness to assess the impact of the spraying on local fauna, such as indigenous pollinators, soil bacteria, fungi, or birds.

The resulting non-knowledge served a strategic purpose as it deprived citizens of information they could use to challenge the government's portrayal of the pesticide as being harmless. Moreover, the absence of knowledge enhanced the effectiveness of other tactics, one of which was appeals to ignorance. As relayed in the last chapter, in trying to portray the pesticide as safe, government officials often claimed there was no evidence suggesting the pesticide was harmful to humans. The effectiveness of such claims is significantly enhanced if no public health research is carried out to systematically assess a product's short- and long-term impacts on human health.

Delaying the Commissioning of Health Impact Research and Circumscribing Its Aims

During the spraying operation's first-year community leaders noted that residents were experiencing health problems linked to the spraying and that the government was failing to assess health effects from the spraying (Green Party 2002a; WASP 2002a, 2002b). Besides publicly denouncing the situation, Dr. Meriel Watts, who was on the PAM Community Advisory Group and was considered the country's foremost pesticides critic, carried out a critical analysis of the Auckland District Health Board's (ADHB) Health Risk Assessment for Foray 48B. The report highlighted several limitations with ADHB's health risk assessment, including the inattention to neurological effects, the failure to adequately characterize the pesticide exposure would be subjected to, an underestimation of inhalation exposure, the systematic discounting of resident reports, the failure to consider the synergistic effects of combining various chemicals, and the twisted logic on which they based their conclusion that the spray was safe for humans. The release of this report in January 2003 created pressure on the Ministry of Health to actually assess the spraying's impact on humans.

Besides Watts's endeavor, the WASP activist group and Hana Blackmore (also on the PAM Community Advisory Group) start collected citizen health experiences, which resulted in Hana Blackmore's February 2003 report. The report documented that over 300 residents suffered respiratory, neurological, and/or digestive problems that they attributed to the spraying operation. In turn, the publicly disseminated report increased the pressure on the health ministry to assess the pesticide's human health effects. After independent reviewers certified the Blackmore report's methodology and analysis, the Ministry of Health finally relented and announced it would consult the public about the spray operation's health impacts (Office of the Ombudsman 2007; Phillips 2003). Towards that end, in March 2003 the Ministry commissioned researchers from the Wellington School of Medicine and Health Services to receive, collect, and summarize reports from the public, community groups, and public health services on the health concerns, symptoms, and effects associated with Foray 48B aerial spraying program (Office of the Ombudsman 2007).

Although commissioning this study might appear to have been a gesture of goodwill towards the community, the action came very late in the process as it came in the 16th month of the spraying operation, which was a month before the extended operation was supposed to end. Moreover, as emphasized by the Ombudsman, most of the initially planned heavy spraying had already taken place *before* the Ministry commissioned the research (Office of the Ombudsman 2007). So, the research was commissioned much too late to prevent the majority of harm that might be caused by the spraying operation.

On the other hand, commissioning the research at such a late date provided a strategic benefit for Ministry of Agriculture and Forestry (MAF) as the absence of alarming information helped maintain the carefully constructed public perceptions about the pesticide's safety. If the research had been commissioned prior to the heavy spraying, there would have been pressure to carry out the research before that spraying could proceed. Moreover, the results could have solidified the public's complaints against the pesticide or perhaps even revealed that the pesticide was more harmful than the public was originally led to believe, both of which would have undermined the perception of pesticide safety the government had carefully cultivated, as well as the public's support for the spraying operation and the government's authority. This underscores that the timing of knowledge production can be as consequential as whether or not the research is actually commissioned or completed.

Another problem with the commissioned research is that the Ministry significantly circumscribed the research team's purview, explicitly forbidding them to conduct research that would specify the number of people impacted by the spraying or identify any causal link between symptoms and health effects (Office of the Ombudsman 2007). Rather, the researchers were limited to cataloguing the health complaints related to the spraying, reviewing the existing scientific knowledge relevant to these health concerns, and recommending scientifically robust further study (*Ibid.*). In turn, while their findings confirmed the complaints that emerged from the community studies, they did not illuminate the number of people impacted by the spraying. In essence, they ended up reproducing the knowledge the community had already produced, rather than building on that knowledge by establishing the extent of the harms, which could have created significant political discomfort for the government. This underscores that those commissioning research can circumscribe its scope in ways that help attenuate its potential impact on safety perceptions.

Delaying the Production and Release
of Uncomfortable Knowledge

Another government strategy was to drag out the production and dissemination of knowledge. Despite the very late commitment to documenting health problems, Ministry of Health officials displayed no urgency in obtaining the community's input. For example, although they announced public consultations would start in April 2003, they set September 30th, 2003, as the report submission date, which was well *after* the extended spray operation's original May 2003 end date. In itself, the timetable suggests that while the Ministry may have been open to having the public share their experiences with researchers, they did not want such knowledge to emerge while pesticide spraying operations were still under way.

In addition to setting a late timetable, knowledge production was delayed in other ways. First, although researchers submitted a complete report on December 17th and hoped to immediately start the peer review process, Ministry officials delayed starting the review process until *after* the New Year (Office of the Ombudsman 2007). Second, after receiving the December report and consulting with MAF, Ministry of Health officials insisted on increasing reviewers, so that the original number was nearly doubled, which increased opportunities for delays (*Ibid.*). Third, Ministry officials were slow to comment on the December draft, sending their comments on February 5th, 2004, instead of the agreed-upon January 30th deadline. Additionally, while the researchers quickly responded to the feedback and expected the final draft to be released by February 23rd, the Ministry, after more consultation with MAF, contended the report needed another major review, which the researchers refused to comply with. Consequently, while the original completion date was September 2003, by mid-April 2004 the report had yet to be publicly released, and government officials were considering to never release it (*Ibid.*). In turn, these delays led a frustrated researcher to leak a draft of the report to the *Wellington Sunday Star Times* on April 17, 2004.

Delaying the report's release was quite advantageous to MAF as the report corroborated the findings from the Blackmore (2003) community health monitoring study. In particular, the researchers reported that those exposed to the pesticide spraying experienced irritant symptoms (i.e. sore throat, headache, and blocked nose), respiratory symptoms (chest tightness, asthma exacerbation, and cough), gastrointestinal symptoms (diarrhea, vomiting, and stomach cramps), flu-like symptoms (fever, malaise, and swollen glands), and skin rash (Hales et al. 2004). Moreover, they reported that those with existing conditions (such as asthma and hay fever) often reported an aggravation in their condition after being exposed to the spraying. Releasing the information at an earlier date would have undermined the perceptions of safety government officials had so carefully cultivated over the previous two years. In turn, this would have eroded public support for the original spraying operation, as well as the government's decision to extend the PAM spraying operation by yet another year (from May 2003 to May 2004) and its decision to pursue a new urban aerial spraying operation in Hamilton (in October and November 2003), following the discovery of one Asian Gypsy Moth in the city (Office of the Ombudsman 2007).

Although some might argue the delays are to be expected with research endeavors, significant evidence suggests the delays in this case were intentional and precipitated by MAF's influence over Ministry of Health officials. For example, in his May 14, 2004, e-mail to the Ombudsman, Simon Hales (i.e. the study's principal investigator) reported:

> Our scientific concerns (in the form of a draft literature review) were made available to MoH in September 2003, and again in more complete form in December, with a recommendation that they be passed on to the Minister. From that time on, it is my perception that MoH began dragging their feet. The number of proposed reviewers was approximately doubled, after consultation between MoH and MAF. MoH then requested that the review be delayed until after the New Year.
>
> (Office of the Ombudsman 2007: 99)

Moreover, after receiving the February draft, the Ministry shared it with MAF officials, who expressed serious concerns with its content and suggested it undergo a major review (*Ibid.*). This precipitated a sharp disagreement with the researchers, who communicated to the Ministry:

> Your intention to seek external review is for political, not scientific, reasons. We do not wish to be involved in your political conflict with MAF. As clearly stated in previous e-mails, we do not intend to revise the report in response to a further round of peer review, at least not as part of the current contract with Health.
>
> (Office of the Ombudsman 2007: 98)

The researchers were vindicated the following month, when a Green Party "Freedom of Information Act" request revealed MAF and MoH bureaucrats had conspired to delay the report's release. Specifically, when MAF officials objected to publicizing the report, on March 4th (which was two weeks *after* the anticipated February 23rd release date), Sally Gilbert (the Environment Team Leader at the Public Health Directorate of the Ministry of Health) informed MAF of delaying tactics at their disposal: (1) "Refuse to accept the report unless it is satisfactory"; (2) "commissioning another analysis and report on the raw data"; and (3) "releasing the report together with an MoH analysis and critique as part of a communications strategy" (Green Party 2004).

This case illuminates a range of tactics that government officials use to manage the production of uncomfortable knowledge so that it has a minimal impact on public support for spraying operations. It behooves social scientists to study this topic more closely. Not only do we need to study the factors mediating whether or not such research is carried out, we also need to understand the factors that mediate *when* such research gets commissioned, the timetable for producing it, who determines the project's scope, how the production of knowledge can get delayed, the timetable for the information's release, and whether the information ever gets released.

Neutralizing Uncomfortable Knowledge

While failing to produce and release potentially uncomfortable knowledge is a powerful means of preserving safety perceptions, when such knowledge exists, government agents and other social forces neutralize it through an array of strategies, including suppression, omission, dismissal, denial, downplaying, and diversion.

Neutralizing Uncomfortable Knowledge through Suppression

Suppression is a particularly powerful strategy and, as Galison (2008) points out, a powerful tactic to conceal information from the public is to classify it as "confidential".

The New Zealand government employed suppression to conceal most of the pesticide's ingredients. While the government identified Btk as Foray 48B's active ingredient, it refused to publicly disclose any of its synthetic ingredients, citing a confidentiality agreement it had signed to protect the manufacturer's "trade secrets." New Zealanders eventually discovered the identity of three ingredients (hydrochloric acid, benzoic acid, and propylene glycol) in May 2003 (Kedgley 2003). However, that discovery came quite late in the process, as the information was not revealed until 17 months of spraying had occurred, which, coincidentally, was the month the extended spraying operation was originally scheduled to end (the campaign was later extended due to the unexpected discovery of more moths). Moreover, the information only emerged because Dr. Meriel Watts and Hana Blackmore sought it through Freedom of Information Act requests and their discovery was then publicly disclosed by the Green Party (*Ibid.*).

Suppressing the identity of the chemicals aided the government because some were associated with health concerns, which, if revealed, could have undermined perceptions that Foray 48B was safe (see Table 8.1). For example, hydrochloric acid is known to cause choking and inflammation of the respiratory tract, with medical research also linking it to asthma, lung cancer, and bronchitis (Brunekeef & Holgate 2002; CHE 2019a; Kedgley 2003; Vizcaya et al. 2011). Regarding benzoic acid, while it is of low concern when consumed in food, except for those who are allergic to it, Watts (2003, 1) emphasizes there is "no known safe level of exposure by inhalation." As for propylene glycol, there is evidence linking it to hearing loss, skin irritation, intestinal damage, depression, and affecting the nervous system of children (CHE 2019b; Kedgley 2003).

Table 8.1 Chemicals Used in Various Foray 48B Formulations

Compound	CAS Number	Potential Health Problems Associated with the Compound
1,5-Hexanediene-3,4-diol, 2,5-dimethyl	4723-10-8	
1-Propanesulfonyl chloride	10147-36-1	Inhalation of this product "is extremely destructive to the tissue of the mucous membranes and upper respiratory tract. Symptoms may include coughing and shortness of breath, and it may cause headaches and nausea" (European Chemicals Agency 2020a). Additionally, ingestion "may cause burns in mouth and throat," skin contact "causes severe skin burns," and eye contact is corrosive, causing stinging, tearing, and redness (*Ibid.*).
4-Acetyloxy-2-Butanone	10150-87-5	Causes skin irritation, causes serious eye irritation, and may cause respiratory irritation (European Chemicals Agency 2020b)
2-Methyl-2,3-pentanediol	7795-80-44	
2,4-Hexadienedioic acid	505-70-4	Suspected to be hazardous to aquatic life and "suspected skin sensitiser" (European Chemicals Agency 2020c)
2-Heptanone,3-hydroxy-3-methyl	13757-91-0	
2-Hydroxy pyridine	142-08-5	"This substance is toxic if swallowed, causes serious eye irritation, causes skin irritation and may cause respiratory irritation" (European Chemicals Agency 2020d).
5-Hexen-2-one, 5-methyl	3240-09-3	"This substance is a flammable liquid and vapour, is harmful if swallowed and causes serious eye damage" (European Chemicals Agency 2020e).
Acetic acid, 2-propenyl ester	591-87-7	Glacial acetic acid is toxic to humans and animals by inhalation and skin contact. Humans exposed to 1,000 ppm for a few minutes may suffer eye and respiratory tract irritation (Patnaik 2007)
Acetic acid, anhydride	108-24-7	"This substance causes severe skin burns and eye damage, is a flammable liquid and vapour, is harmful if swallowed and is harmful if inhaled" (European Chemicals Agency 2020f)
Benzoic acid[a]	65-85-0	Asthma, and irritates the skin and eyes (Brunekeef & Holgate 2002; Vizcaya et al. 2011). Toxic symptoms include somnolence, respiratory depression, and gastrointestinal disorder (Patnaik 2007). "This substance causes damage to organs through prolonged or repeated exposure, causes serious eye damage and causes skin irritation" (European Chemicals Agency 2020g)
Butylated hydroxy toluene	128-37-0	"This this substance is very toxic to aquatic life with long lasting effects" and is under assessment for endocrine disruption (European Chemicals Agency 2020h)

Cyclohexasiloxane, dodecamethyl	540-76-6	
decamethyl Cyclopentasiloxane,	541-02-6	This substance is "persistent, bio-accumulative and toxic" (European Chemicals Agency 2020i)
Cyclotetrasiloxane, octamethyl	556-67-2	This substance is "persistent, bio-accumulative and toxic." Additionally, it "is a flammable liquid and vapour, is suspected of damaging fertility or the unborn child and may cause long lasting harmful effects to aquatic life" (European Chemicals Agency 2020j)
Cyclotrisiloxane, hexamethyl	541-05-9	This "substance causes serious eye irritation, may cause respiratory irritation and causes skin irritation" (European Chemicals Agency 2020k)
Disiloxane derivative	18420-09-2	"This substance is a flammable liquid and vapour, causes serious eye irritation, causes skin irritation and may cause respiratory irritation" (European Chemicals Agency 2020l)
Hydrochloric acid[a]	7647-01-0	"This substance causes severe skin burns and eye damage, is toxic if inhaled, may damage fertility or the unborn child, causes serious eye damage, may cause damage to organs through prolonged or repeated exposure, may be corrosive to metals and may cause respiratory irritation" (European Chemicals Agency 2020m). Additionally, inhalation can cause choking and inflammation of the respiratory tract, and the substance is linked to asthma, lung cancer and bronchitis (CHE 2019a; Brunekeef & Holgate 2002; Patnaik 2007; Vizcaya et al. 2011).
Methyl paraben	99-76-3	This substance is toxic to aquatic life and is being assessed for endocrine disruption. (European Chemicals Agency 2020n). Also, it is linked to: (1) skin irritation and allergy; and (2) increases in breast cancer tumor proliferation (Ishiwatari et al. 2006; Lillo et al. 2016; Matwiejczuk et al. 2020).
Penta siloxane, dodecamethyl	141-63-9	
Phenyl amine R silane derivative	10538-85-9	
Trimethyl phosphine	594-09-2	"This substance is a highly flammable liquid and vapour, causes serious eye irritation, causes skin irritation and may cause respiratory irritation" (European Chemicals Agency 2020o). Also, preliminary evidence links phosphine to cataracts, chronic renal disease, heart attacks, and peripheral neuropathy (CHE 2019d)
Phophoric acid	7664-38-2	(1) Dermatological exposure can irritate skin and mucous membranes; (2) its vapors can cause coughing and throat irritation (Patnaik 2007)
Potassium phosphate	7778-53-2	"This substance causes serious eye damage and may cause respiratory irritation" (European Chemicals Agency 2020p)

Compound	CAS Number	Potential Health Problems Associated with the Compound
Propylene glycol[a]	57-55-6	Linked to hearing loss, skin irritation, intestinal damage, depression, and has been found to affect children's central nervous systems (CHE 2019b; Kedgley 2003)
Sodium hydroxide	1310-73-2	"This substance causes severe skin burns and eye damage" (European Chemicals Agency 2020q). Additionally, it is: (1) Severely corrosive to eyes, skin, mucous membranes and digestive systems; (2) breathing sodium hydroxide dust or mist leads in mild cases to irritation of the mucous membranes of the nose… and in severe cases to damage of the upper respiratory tract" (Harte et al. 1991)
Sulfuric acid	7664-93-9	(1) Can cause severe skin burns and permanent vision loss; (2) inhaling it as a mist or vapors can produce coughing and significant bronchial constriction; and (3) chronic exposure can produce bronchitis, conjunctivitis, skin lesions, and erosion of teeth; (4) cause laryngeal cancer (CHE 2019c; Patnaik 2007)
Sydnone, 3-phenylmethyl	16844-42-1	
Thietane	287-27-4	"This substance is a highly flammable liquid and vapour and is harmful if swallowed" (European Chemicals Agency 2020r)
Trisiloxane	3555-47-3	

[a] Chemical was confirmed to be in the formulation used in the PAM operation.

The problem's significance grows when we consider those were only 3 of the 19–30 chemicals typically found in Foray 48B formulations. For instance, the Canadian formulation has 19 chemicals, the identity of which the Canadian government also withheld from the public, though that information eventually saw the light of day when University of British Columbia scientists reverse-engineered the pesticide and publicly disclosed their findings (van Netten et al. 2000). Chemicals in the latter formulation were also associated with health concerns as sulfuric acid causes laryngeal cancer and phosphine is associated with the onset of hepatitis, pulmonary edema, seizures, cataracts, chronic renal disease, heart attacks, and peripheral neuropathy (CHE 2019c, 2019d).

Concealing the identity of a pesticide's ingredients is a powerful tactic to stymie critiques against that pesticide. If citizens know a pesticide's ingredients, they can track down the published research associated with each chemical and use that knowledge to predict the pesticide's potential health impacts, which could seriously undermine government communication efforts. Conversely, suppressing that information meant that Aucklanders could not identify what they were being exposed to and what harm it could do. In turn, this socially produced ignorance helped attenuate risk perceptions and helped maintain public support for the spraying operation. While it is true spray victims would have been keenly aware of the linkages between their health problems and the pesticide spraying, the socially produced ignorance about the pesticide ingredients would have made it more difficult to persuasively explain to others how the pesticide was contributing to their health problems, thereby also stymying the spread of concern about and opposition to the spraying.

Beyond helping attenuate risk perceptions in its own right, concealing ingredients enhances the effectiveness of other tactics, such as portraying a pesticide as being harmless or dismissing complaints against it. If the public cannot identify the individual chemicals in a pesticide, it makes it much more difficult to challenge claims the pesticide is harmless or that its effects will be insignificant.

Neutralizing Uncomfortable Knowledge through Omission

When uncomfortable knowledge about pesticide spraying cannot be suppressed, such knowledge becomes a threat to the spraying's proponents, who will pivot to other neutralization strategies, such as omitting that knowledge from communications about the product.

In the case of Foray 48B, there were many examples of this tactic being deployed. First, government officials completely omitted any mention of health problems when they claimed that Foray 48B is harmless to humans. This even included omitting the health safety warning provided by the manufacturer's Safety Data Sheet, which emphasized there would be issues for at least a minority of those exposed to the pesticide (Novo Nordisk 1991; Valent Biosciences 2009).

Another example was the government's claim that the pesticide's main ingredient (i.e. Btk) is harmless to humans, which also summarily ignored contradictory evidence. While MAF asserted Btk could only become active in the alkaline gut of caterpillars, previous research had revealed that Btk spores can survive and remain viable in humans months after exposure and that this was even true of those with healthy immune systems (Bernstein et al. 1999; Valadares et al. 2001). Other research revealed that Bt in commercial pesticides consistently produces an enterotoxin that can cause food poisoning symptoms, including nausea and vomiting (Damgaard 1995; Tayabali & Seligy 2000).

Omission was also at play with government claims the Btk bacteria only affects caterpillars, which ignored evidence that mammals were affected by Btk. For instance, Jenkins (1992, as cited in Swadener 1994) found that lab rats experienced respiratory depression when exposed to air containing Btk spores and that rabbits exposed to Bt experienced irritation to the skin and eyes. Moreover, during the PAM spraying campaign there were over 20 reports of cats being unwell following sprayings, with typical symptoms including vomiting, lack of hunger, infected eyes, and skin allergies (Blackmore

2003). There were also reports of dogs being affected, with the most common symptom being diarrhea, followed by eye and skin problems (*Ibid.*). This information was never reflected in government pronouncements about the Btk bacteria or the Foray 48B pesticide.

Carrying out such omissions considerably benefitted MAF as communicating the health concerns would have raised concerns about the pesticide and, in turn, could have eroded public support for the spraying operation.

Dismissing Uncomfortable Knowledge through Denial

Another neutralizing strategy MAF employed was dismissal. Rayner (2012) defines dismissal as engaging with uncomfortable knowledge, in order to rebut it. There are many dismissal tactics, with one being to outright deny the charges laid against a product, as the lead, chemical, and tobacco manufacturers did successfully for decades (Markowitz & Rosner 2002; Moyers et al. 2002; Proctor 2008).

As this pertains to the PAM operation, there seems to have been a systematic attempt at denial, as health workers fielding calls from sick residents consistently denied claims that the health problems were due to the spraying and instead attributed the symptoms to psychosomatic or other causes (Goven et al. 2007; Office of the Ombudsman 2007). This point is underscored by the Commissioners of the People's Inquiry (Goven et al. 2007), who reported that the Aeraqua[1] health workers who fielded calls routinely trivialized and psychologized symptoms reported by residents:

> Attempts to access health support directly through the Health Service phone line proved distressing for many of those who provided testimony. We heard repeatedly of people being told during these telephone conversations with Aeraqua nurses (that is, with no physical examination) that their symptoms, including skin rashes and blistered skin, nose ulcers, and respiratory difficulties, could not possibly be caused by the spray. Even more distressing for the callers, other explanations were offered for the symptoms, without any examination or evidence. Examples we heard included menopause (for symptoms of allergy), heredity (for nose ulcer), and hysteria or psychosomatic reaction (for a wide range of symptoms).
>
> (p. 13)

Not only was it demeaning for the residents to have their situation trivialized, it also meant they could not access services that the government had promised to those who were affected by the spraying.

Even if residents made it through the initial phone screening, many faced similar attitudes from the Aeraqua doctors, who locals referred to as "MAF doctors." According to the aforementioned Commissioners, many residents reported the doctors also exhibited a tendency to attribute symptoms to any cause but the spray,

> and that often there seemed to be a refusal to consider the possibility that the spray may be aggravating a pre-existing condition (such as asthma). People with respiratory difficulties were told that the fact that they had asthma before the spraying began meant that their current difficulties could have nothing to do with the spray.
>
> (Goven et al. 2007, p. 13)

These points were further underscored by testimony that several residents offered during the Public Inquiry:

> The MAF doctor was no charge I recall, but it was a very unsatisfactory visit as she was trying to convince me that the reaction [irritated, inflamed eyes coinciding with spray events] was caused by other things, and that the spray was safe and couldn't possibly have this effect.
>
> (Goven et al. 2007, p. 14)

> The MAF medical [service] tried to convince me that the face rash was caused by coming into contact with the Painted Apple Moth.
>
> (Goven et al. 2007, p. 14)

> I was sent to two [Aeraqua doctors] and they said it wasn't the spray that caused my illness it was my age. I gave both of them permission to get my medical files from my Dr but they did not do so.
>
> (Goven et al. 2007, p. 14)

> At the time MAF was offering free consultations with doctors chosen by them for people who claimed to be affected by the spray. I went to one of these doctors, who told me my symptoms were not related to the effects of the spray. I did not accept this and went to the doctor I normally consult… He confirmed my suspicions that the symptoms were due to spray exposure and issued me a certificate which specified that I should be evacuated from the area when the spraying was due to take place. The symptoms have not recurred at any time since.
>
> (Goven et al. 2007, p. 14)

> The response from the official medical officers doctors and nurses employed by Aeraqua was one of scepticism and refusal to validate that any of the symptoms were spray related, even though they must have had hundreds of reports. One nurse used the post-modern language of subjectivism (… everybody has their own reality).
>
> (Goven et al. 2007, p. 14)

Having healthcare workers–who are in positions of considerable authority– consistently deny the relationship between symptoms and the spraying, psychologize the symptoms, and ridicule the patients for holding their beliefs, would have led many victims to develop self-doubts about the links they were noticing. In turn, this would have deterred many from pressing on with their claims or sharing them with others, which would also stymied the spread of opposition against the spraying operation.

Dismissal by Downplaying the Significance of Incriminating Knowledge

Another dismissal tactic is to downplay the significance of uncomfortable knowledge. One way to downplay uncomfortable knowledge is to suggest the problem represents insignificant risk to those affected by it. This too was a tactic employed by New Zealand government officials. One manifestation was the government's tendency to acknowledge some symptoms but to refuse to characterize them as health problems, as illustrated by the following MAF statement: "While the committee found that while there were minor respiratory irritations at the time, there was no evidence of health problems caused by the spray" (MAF 2001b). If respiratory problems are not considered a health problem, it begs the question of how they defined "health problems."

Another way to downplay evidence is to portray incriminating findings as only relevant for a small subset of people, a tactic pharmaceutical manufacturers routinely employ in their US television ads. New Zealand officials deployed this tactic by suggesting only a tiny fraction of the population would be affected by the spray. For example, at the end of the first year of the PAM operation the Biosecurity Minister acknowledged there were some concerns with the Foray 48B pesticide, but downplayed those concerns by stipulating that only "about 5 per cent of the population were allergic to the Btk spray" (Green Party 2002c; Sutton 2002). Given the large number of people who were repeatedly exposed to the pesticide [between 200,000 and 300,000 (Office of the Ombudsman 2007)], even 5% is a significant number as it equates to between 10,000 and 15,000 people. Moreover, these figures do not include the numerous people who would be affected by any of the 20–30 synthetic ingredients added to the pesticide.

A third way MAF downplayed incriminating evidence was to suggest the only people who would be affected were those with pre-existing conditions, which represented a small portion of the population. For example, three months after spraying began a MAF (2002b) press release stated: "Most people will not be affected by the spray, unless they have specific allergies." What this leaves unsaid is that even if the only people affected were those with specific allergies, that could add up to a significant number of people. For example, the Wellington School report (Hales et al. 2004) indicated that asthma is one of the conditions that was worsened by the spraying, and asthma is one of the conditions that afflicts a significant number of New Zealanders (12.5% of adults and 14.3% of children (Health Navigator New Zealand n.d.). In turn, 12.5% of 200,000 Aucklanders equates to 25,000 people. Even if these were the only ones affected by the spray, this would be a considerable number.

Still another way they downplayed evidence was to characterize any risks to human health as being quite low. For example, in October 2002, when the government was significantly expanding the spray zone (from 2,000 to over 20,000 acres), a MAF press release communicated:

> MAF would like to assure people who live and work in the painted apple moth zone that the spray has been chosen because it is proven to be an effective way of killing caterpillars yet with low risks to human health.
>
> (MAF 2002c)

Downplaying the significance of uncomfortable knowledge is arguably a more powerful neutralization strategy than denial. Outright denial is a heavy-handed rhetorical tactic that comes across as disingenuous. Downplaying, on the other hand, is a more sophisticated rhetorical ploy, as acknowledging that some people will get hurt conveys a reasonableness that builds credibility among those who are not directly affected, who, in turn, will be more likely to lend their support to spraying operations.

Neutralizing Uncomfortable Knowledge through Diversion

Another neutralization strategy is to pursue communication activities that will divert attention from the health concerns associated with the product. An example of this tactic is the great lengths tobacco manufacturers took to divert the public's gaze from the relationship between tobacco and cancer, which included funding research to redirect attention to other potential causes of lung cancer and hiring historians to create a positive narrative of the tobacco industry (Proctor 2008).

Regarding the PAM spraying operation, MAF deployed diversion in three ways: (1) they stressed Btk's "natural" origins; (2) they associated Foray 48B with organic agriculture; and (3) they emphasized that the synthetic chemicals were approved food additives. Each of these served to divert attention from the pesticide's potential health impacts.

Prior to and during the first year of the spraying operation, MAF sought to portray Btk as "natural" through statements like:

> Bacillus thuringiensis var. kurstaki (Btk) is a bacterium that occurs naturally in soil, foliage, water and air in most countries in the world, including New Zealand.
>
> (MAF 2001a)

> Btk is found naturally in soil, air and water.
>
> (MAF 2002a)

The active ingredient in the Foray spray is a naturally-occurring soil bacterium called Bacillus thuringiensis kurstaki or Btk for short.

(MAF 2002c)

This same organic, naturally-occurring spray was used in the eradication of the white spotted tussock moth, and there were no serious ill effects then.

(MAF 2002b)

Portraying Btk as "natural" was a powerful way to connote safety as people tend to view natural products as being safer than synthetic pesticides. This is particularly true in an era where people have become increasingly concerned about human health harms caused by synthetic chemicals.

However, there are numerous problems with associating Btk's "natural" origins with safety. First, "natural" does not mean something is automatically safe for humans. The planet is full of "natural" bacteria that are quite harmful to humans, including *Bacillus cereus* (which can induce food poisoning) and *Bacillus anthracis* (also known as anthrax, which can cause skin sores, vomiting, and shock), both of which are related to Btk. Second, while the natural form of Btk might be relatively benign, the strain used in sprays has been engineered to be at least six times more potent (Sanahuga et al. 2011; Swadener 1994). Thus, it was misleading, and arguably negligent, for MAF to suggest the Btk in Foray 48B was as benign as the natural strain found in gardens.

A third problem is that while people can be exposed to low levels of Btk in their garden soil or through ingesting organically produced food, it is an entirely different matter to expose people to the bacterium in aerosol form, which facilitates the inhalation of the spores into the nasopharynx and lungs. The matter takes on greater weight when we consider that instead of receiving one exposure per day (which is what the government's health risk assessment was premised on) many residents were repeatedly exposed on spray days, due to the multiple sprayings administered each day, as well as legacy exposure in their homes, which were increasingly contaminated by the spraying. While epidemiological studies of aerosolized Btk had yet to demonstrate health effects, these studies had important limitations, including small sample sizes and "biased assessment of health effects, potential or actual exposure of control groups, and limited duration of follow-up" (Hales 2004).

Another diversion tactic was linking the pesticide to organic agriculture, through statements like:

From late November 2001 the organic pesticide Btk will be used again when the next phase of the programme to attempt to eradicate painted apple moth commences with targeted aerial spraying.

(MAF 2001b)

Some Btk products are approved organic pest control agents.

(MAF 2002c)

Although Btk is found naturally in the soil and is a commonly used agricultural insecticide in organic production...

(MAF 2001b)

New Zealand organic growers have been using Btk spray since 1984.

(MAF 2001a)

Such carefully crafted statements powerfully connote safety, due to the fact citizens typically associate organic agriculture with safety and health. However, this rhetorical ploy is also problematic. Even though Btk is a component in *some* organic sprays, Foray 48B has never been considered an organic product due to the synthetic adjuvant chemicals that are added as "inert" ingredients. Underscoring this point is that in 2003 organic certifiers warned farmers they would lose their organic certification if they sprayed their produce with Foray 48B (New Zealand Herald 2003).

The third diversion tactic was to represent the synthetic chemicals as approved food additives. For example, in one press release MAF officials made the following statement: "While the spray also contains food residues, preservatives, an acidity regulator, an alcohol and a sugar-like substance as a stabiliser, these ingredients are approved food additives" (MAF 2001a).

This is another way to powerfully connote safety as things we ingest are commonly perceived to be safe. However, this too was misleading. While a chemical can be approved for ingestion, it does not logically follow that it is safe to inhale it or to expose one's eyes to it. Indeed, while they may be classified as food additives, each of the three uncovered chemicals in the Foray 48B formulation (benzoic acid, hydrochloric acid, and propylene glycol) are associated with health problems. As previously mentioned, research has found that exposure to benzoic acid irritates the skin and eyes and is linked to asthma (Brunekeef & Holgate 2002; Vizcaya et al. 2011). Moreover, as Watts (2003) emphasizes "whilst benzoic acid is regarded as being of low toxicity when ingested, except to those allergic to it, there is no known safe level of exposure by inhalation" (p. 1). Additionally, hydrochloric acid is associated with inflammation of the respiratory tract, choking, asthma, bronchitis, and lung cancer (CHE 2019a; Brunekeef & Holgate 2002; Patnaik 2007; Vizcaya et al. 2011). As for propylene glycol, it is linked to hearing loss, skin irritation, intestinal damage, and depression, and has been found to affect children's central nervous systems (CHE 2019b; Kedgley 2003).

Diversion is another powerful neutralization strategy as it can effectively distract attention from concerning information associated with the pesticide's ingredients, which can help perpetuate false perceptions that the pesticide is safe.

Neutralizing Potential Sources of Uncomfortable Knowledge

Aside from stalling the production of uncomfortable knowledge and deploying rhetorical tactics to neutralize existing knowledge, government officials also stymied potential sources of uncomfortable knowledge (including the Community Advisory Group, frontline medical workers, and the Ministry of Health) from disseminating such knowledge.

Reducing the Community Advisory Group's Effectiveness

The Community Advisory Group (CAG) was an important potential source of health-related information as its membership contained a doctor and at least two people (i.e. Hana Blackmore and Dr. Meriel Watts) who had lived through the 1996–1997 Project Ever Green eradication operation and who were therefore quite familiar with the pesticide's potential health problems. Additionally, part of the group's mandate was to listen to and communicate community concerns about the spraying to MAF, and, when MAF failed to listen to those concerns, the group showed a willingness to take those issues to the media (PAM CAG 2002a, 2002b).

MAF used several tactics to curtail the group's potential effectiveness. The first was waiting until September 2001 to establish such a group, which was 27 months after PAM was first discovered (Walsh 2003). This made it harder for the community to have much say in MAF's initial efforts to eradicate the moth. While it is true that activist groups could have formed and voiced their opinions before CAG's establishment, they would not have had as much legitimacy with the public as a group that was formally recognized by MAF.

Once advisory groups are formed, one tactic to control them is to appoint government-friendly people to leadership positions, which is something MAF attempted when it first established CAG (personal communication with advisory group member). However, this attempt failed as the first meeting was very well attended by local residents, who lobbied hard regarding the group membership and leadership.

One outcome of the community's lobbying is that it forced MAF to accept Hana Blackmore and Meriel Watts on the advisory group. This was significant because both had lived through the 1996–1997 eradication operation, were quite aware of how eradication operations could impact communities, and had proven to be quite critical of pesticide spraying. Additionally, locals pressured MAF to accept Kubi Witten-Hannah (a school teacher with deep roots in the community) as the leader of the advisory group. These outcomes led to a Community Advisory Group that was quite outspoken and, when MAF proved unwilling to engage in good faith, showed itself willing and adept at using the media to pressure MAF, which included calling for the PAM operation director's resignation in December 2001 (PAM CAG 2001).

MAF's third strategy to curtail CAG's effectiveness was to disband it and reconstitute it with government-friendly members. The disbanding occurred in November 2002, a few months after the arrival of Robert Isbister, the newly appointed general manager of PAM operations (Walsh 2003). In explaining his decision, Isbister stated, "I was told by MAF officials that this group had lost the plot and I was to disengage. I wasn't to get involved with them…meaningful dialogue was not possible" (as cited in Walsh 2003). On the other hand, Dr. Meriel Watts (who had also been on the Community Advisory Group during the 1996–1997 Project Ever Green) was quite critical of MAF for their lack of real engagement with the community: "Their idea of consultation is using a community group to smooth everything over – it's a charade of consultation" (as cited in Walsh 2003).

Although Isbister had originally promised to reconstitute the group by December 2002, he did not actually reconstitute the group until February 27, 2003, which meant the community went nearly four months without a formal means of communicating with MAF (Green Party 2002c; Office of the Ombudsman 2007). Moreover, when he reconstituted it he stacked it with government-friendly people and appointed himself the group's chair, thereby ensuring complete control over the group (Green Party 2002c; Walsh 2003).

Regarding the strategy's impact, while MAF disbanded the original group, the effectiveness of that strategy was reduced by the fact the members retained the support of Bob Harvey (Waitekere City mayor) and the full Waitakere City Council, who provided the group with meeting space and other operational resources (PAM CAG 2002c; personal communication with group member). In turn, the members continued speaking on behalf of the community, while also continuing to publicly raise concerns about the spraying and MAF's misconduct (PAM CAG 2003a, 2003b; PAM Community Coalition 2002). Additionally, two of their members (Meriel Watts and Hana Blackmore) produced health reports that undermined MAF's narrative about the pesticide's harmlessness, which pressured the Ministry of Health to at least give the appearance they were concerned about the pesticide's effects on people, by commissioning research to investigate the human health impact of the spraying (Office of the Ombudsman 2007). Furthermore, the continued support enjoyed by the original group played an instrumental role in the development of the People's Inquiry, which created a forum for residents to publicly share their experiences and resulted in a report produced by the Commissioners of the Inquiry (Goven et al. 2007).

Handcuffing the Medical Workers

Another potentially important source of uncomfortable knowledge were the doctors and nurses who were engaging with residents in the spray zone who were experiencing health problems. These medical workers were pivotal to MAF's cause because they were in a position to either affirm or deny their patients' belief that the health problems were linked to the spraying.

To control this situation MAF assumed responsibility of the health services provided to spray zone residents (which were called the PAM Health Services). As part of that process, MAF contracted the medical services to Aeraqua, a private medical group that provided nurses to handle the telephone lines that residents were asked to call to report health problems and doctors to examine and diagnose spray zone residents who complained of health problems.

This arrangement was quite odd. The most logical arrangement would be to have health matters overseen by the Ministry of Health, whose primary objective is to safeguard public health. Conversely, the Ministry of Agriculture and Forestry is not responsible for safeguarding human health. Rather, its main responsibility is protecting and enhancing the profitability of the agricultural and forestry sectors. Thus, the arrangement created a significant conflict of interest, as it was highly unlikely that MAF would ever prioritize human health concerns over their primary mandate of protecting and enhancing forestry and agricultural concerns. More importantly, this arrangement made the medical workers financially beholden to MAF, which incentivized them to avoid saying anything that could undermine the official narrative that the pesticide was harmless. In turn, this helps explain the culture of dismissal that Auckland residents faced when they called on the phone hotline and received in-person assessments by doctors.

Some might be tempted to defend the healthcare workers by arguing they are guided by the Hippocratic Oath to protect patients. However, as social scientists have illuminated, funding arrangements are a powerful way to strait-jacket doctors into supporting the dominant narrative. In her own work on the issue, Barbara Ellen Smith (1981) found that funding arrangements in the coal-mining industry were pivotal to the way doctors treated workers. In particular, she illuminates that when employee collectives or unions funded medical services, doctors were far more likely to attribute symptoms of black lung disease to the unsafe environments miners were working in. However, when the mine owners were funding medical services, company doctors tended to downplay health issues that coal miners were experiencing and accused them of malingering. Moreover, when doctors did acknowledge health issues, they tended to deny that the workplace had anything to do with those symptoms and instead attributed blame to the workers' carelessness or hurtful forms of recreation, such as alcohol consumption.

A similar culture of dismissal manifested itself in the PAM case. As related earlier in this chapter, the People's Inquiry (Blackmore 2020; Goven et al. 2007) revealed that MAF-funded doctors systematically sought to deny the link between the health symptoms and the sprayings. Additionally, as part of that process, they sought to redirect patients to a range of other potential explanatory factors, including heredity, menopause, psychosomatic reactions, the patient's age, and coming into contact with a painted apple moth (Goven et al. 2007). Making the situation even more incredulous to patients is that these pronouncements were made by phone nurses, who drew their conclusions without even seeing the patients (*Ibid.*).

Besides the funding arrangement, another factor that contributed to the culture of dismissal was the conflict of interest created by appointing Dr. Francesca Kelly to direct the PAM Health Services. Dr. Kelly had authored the Foray 48B health risk assessment produced for the 1996–1997 Project Ever Green, which had concluded that the pesticide would be harmless to residents. Additionally, she directed and authored the post-surveillance reports for that operation, which concluded that the spraying had been harmless to the residents. Appointing her to direct the PAM Health Services was a conflict of interest because she would have a vested interest in ignoring or suppressing any evidence that could call into question her previous intellectual work. Moreover, as the director of that operation, she would have had the power to pressure those under her to also ignore such evidence.

Silencing the Ministry of Health

A third group that could have shared uncomfortable knowledge was the Ministry of Health, whose primary responsibility was to protect the public's health. As part of that responsibility, the Ministry was supposed to identify and raise health concerns associated with the proposed pesticide use, see to it that any identified risks are appropriately mitigated, ensure that health risk assessments for the pesticide have exposure scenarios that correspond with how the pesticide will actually be used, have the health risk assessments redone if the citizens' exposure to the pesticide exceeds what was predicted in the original health risk assessment, carry out studies to assess what impact the pesticide spraying is having on local residents, and provide health care to those affected by the spraying.

In carrying out such responsibilities, the Ministry of Health would have been in a position to disseminate knowledge that could be quite inconvenient to pesticide proponents at MAF. However, the Ministry of Health failed to carry out any of the aforementioned tasks. Rather, it supported MAF's spraying operation through a variety of actions and inactions. First, it released a deeply flawed Health Risk Assessment, which failed to consider neurological effects, used exposure scenarios that grossly underestimated the amount of exposure residents would get, underestimated the impact of inhalation exposure, systematically discounted reports provided by residents in previous spraying operations, and failed to consider the synergistic effects of the chemicals in the pesticide (Watts 2003). Second, when it became evident the expansion of the spraying operation would far surpass the exposure scenarios used in the Health Risk Assessment, the Ministry failed to produce a new assessment (Office of the Ombudsman 2007). Nor did the Ministry raise any health concerns about the expanded spraying operation, as underscored by the Ombudsman: "I am informed that the Ministry of Health had no concerns with the proposal to move to expanded aerial operations. Human health does not appear to be mentioned in the Executive Summary of the Minister's paper nor in the Cabinet paper itself" (2007: 41). Fourth, after the spraying had begun the Ministry refused to carry out studies to assess how the spraying was impacting residents (*Ibid.*). Moreover, when public pressure forced them, in the second year of spraying, to commission such research, the Ministry undermined the project by circumscribing its scope and conspiring with MAF to delay the production and release of the research results (*Ibid.*). Lastly, the Ministry allowed MAF to oversee the care provided to spray victims, even though MAF had no expertise in this area and their oversight of the PAM Health Services represented a direct conflict of interest with their primary responsibility to protect and enhance the forestry sector, which it was doing by seeking to eradicate PAM through the use of aerial pesticide spraying. The Ombudsman (2007) captured the situation well when he argued the Health Ministry completely abdicated on their responsibility to protect the public's health.

For an agency charged with protecting the public's health, the Ministry of Health's performance was particularly egregious and begs the question of how such a development could occur. The main factor was that in their handling of the PAM incursion Helen Clark's government opted to use a "whole-of-government" approach, which meant that all other government agencies were subordinated to MAF's objective of eradicating the moth (Office of the Ombudsman 2007). A whole-of-government approach enables a government bureaucracy to channel all its resources towards a single goal, which can significantly increase its chances to accomplish that goal.

However, such an arrangement raises significant and troubling questions. Should a segment of the population ever be put in harm's way to protect industry profits? Even if we accept that "only" 10% were at risk from the spraying, which would to be a very conservative figure, 10% still represents 20,000 people out of the 200,000 who were exposed to the pesticide. On what basis are government officials allowed to treat the health and well-being of these people as acceptable collateral damage, to ensure that an environmentally destructive industry will avoid having its profit maximization *potentially* threatened by a foreign species? Why aren't government officials also considering the monetary impact associated with the harm done to these people, which includes healthcare costs, the loss of productivity and livelihood, the loss of well-being, and the loss of potential service they could be providing to the community? Moreover, if the agency in charge of protecting human health is prevented from doing its job, who will protect the people? Relatedly, if the Ministry of Health will not do its job, why should residents be deprived of any legal recourse? As discussed earlier, MAF officials arranged to alter the Resource Management Act prior to the PAM operation, which gave the agency a legal loophole that enabled it to forge ahead with pesticide spraying, despite concerns locals may have had about it. In such a scenario, where are the democratic checks and balances to ensure that citizens are protected from government overreach and, in particular, overzealous government technocrats that are single-mindedly obsessed with protecting industry interests? Furthermore, this whole case begs us to ask who is really being served by biosecurity operations and whether they are simply another government apparatus to maximize industry profits at the expense of citizens.

Summary

As articulated at the chapter's outset, while portraying pesticides as harmless can help maintain support for a spraying operation, the success of those efforts will be mediated by the pesticide proponents' ability to effectively neutralize uncomfortable knowledge. This chapter illuminates the numerous strategies that government officials deployed to manage the uncomfortable knowledge pertaining to Foray 48B, which included: (1) delaying the commissioning of public health research; (2) circumscribing the parameters of such research; (3) delaying its production and release; (4) neutralizing uncomfortable knowledge by suppression, omission, dismissal, denial, downplaying, and diversion; and (5) hampering key groups that could disseminate uncomfortable knowledge.

The government's manipulation tactics drew the scorn of the Ombudsman, whose report criticized government agencies for both failing to pursue the research that would systematically assess the spraying operation's health impacts and for pursuing a communication campaign that misleadingly reassured the public (Office of the Ombudsman 2007). Regarding the undone science, he judged the problem to be so serious that he recommended the government: (1) fund research to investigate the long-term human health effects of the PAM spraying operation; and (2) forestall pursuing similar urban spraying operations until it is established that Foray 48B has no long-term human health effects (Office of the Ombudsman 2007). It is telling that the New Zealand government never commissioned such a study, nor has it since undertaken any other urban aerial pesticide spraying operation, whether with Foray 48B or any other pesticide.

This chapter underscores that far from being a neutral arbiter on such issues, government agencies actively work to manipulate public perceptions about pesticides, which, in turn, significantly contributes to the acceptance and normalization of pesticide use. Moreover, by illuminating this case, the analysis should prove useful for understanding how governments in other locales manage both uncomfortable knowledge and public support for urban aerial pesticide spraying operations and other pesticide use.

Having said this, while the management of uncomfortable knowledge is central to maintaining public support, we also need to consider how the effectiveness of such activities is mediated by cultural context, which includes the cultural acceptance of pesticide use and the factors that contribute to that acceptance (including pesticide ignorance, an acquiescence to toxicity, and key educational deficiencies). These are issues I address in the next chapter.

Note

1 Aeraqua was the private firm that MAF hired to handle all health services provided to residents who experienced health effects from the spraying. Hiring an outside firm to provide health services (instead of allowing the care to be overseen by the Ministry of Health) enabled MAF to ensure that the advice and care given by these health workers would not undermine MAF's messaging that the pesticide was safe for humans.

References

Bernstein, Leonard, Jonathan Bernstein, Maureen Miller, Sylva Tierzieva, David Bernstein, Zana Lummus, MaryJane Selgrade, Donald Doerfler, and Verner Seligy. 1999. "Immune Responses in Farmer Workers after Exposure to Bacillus thuriengiensis Pesticides." *Environmental Health Perspectives* 107(7): 575–582.

Blackmore, Hana. 2003. *Interim Report of the Community-based Health and Incident Monitoring of the Aerial Spray Programme.* Auckland, New Zealand.

Blackmore, Hana. 2020. *Btk Pesticide Sprays: Adverse Health Impacts and Effects: Cover up or Abdication of Responsibility?* Unpublished report. Auckland, New Zealand.

Brunekeef, Bert and Stephen Holgate. 2002. "Air Pollution and Health." *Lancet* 360(9341): 1233–1242.

CHE (Collaborative on Health and the Environment). 2019a. "Toxicant and Disease Database Entry for 'Hydrochloric Acid'." Accessed February 1, 2022. https://www.healthandenvironment.org/our-work/toxicant-and-disease-database/?show category=&showdisease=&showcontaminant=2651&showcas=&showkeyword=

CHE (Collaborative on Health and the Environment). 2019b. "Toxicant and Disease Database Entry for 'Propylene Glycol'." February 1, 2022. https://www.healthandenvironment.org/our-work/toxicant-and-disease-database/?showcategory=&showdisease=&showcontaminant=2995&showcas=&showkeyword=

CHE (Collaborative on Health and the Environment). 2019c. "Toxicant and Disease Database Entry for 'Sulfuric Acid'." February 1, 2022. https://www.healthandenvironment.org/our-work/toxicant-and-disease-database/?showcategory=&showdisease=&showcontaminant=2798&showcas=&showkeyword=

CHE (The Collaborative on Health and the Environment). 2019d. "Toxicant and Disease Database Entry for 'Phosphine'." Accessed February 1, 2022. https://www.healthandenvironment.org/our-work/toxicant-and-disease-database/?showcategory=&showdisease=&showcontaminant=2364&showcas=&showkeyword=

Damgaard, Per Hyldebrink. 1995. "Diarrhoeal Enterotoxin Production by Strains of *Bacillus thuringiensis* Isolated from Commercial *Bacillus thuringiensis*-based Pesticides." *FEMS Immunology and Medical Microbiology* 12: 245–250.

Déplaude, Marc-Olivier. 2015. "Minimising Dietary Risk: The French Association of Salt Producers and the Manufacturing of Ignorance." *Health Risk & Society* 17(2): 168–183.

Diller, Lawrence. 1999. *Running on Ritalin: A Physician Reflects on Children, Society, and Performance in a Pill.* New York: Bantam Books.

European Chemicals Agency. 2020a. *Entry for "1-Propanesulfonyl Chloride" in the REACH Database.* https://echa.europa.eu/registration-dossier/-/registered-dossier/30670/9

European Chemicals Agency. 2020b. *Entry for "4-Acetyloxy-2-Butanone" in the REACH Database.* https://echa.europa.eu/information-on-chemicals/cl-inventory-database/-/discli/notification-details/154301/608336

European Chemicals Agency. 2020c. *Entry for "2,4-Hexadienedioic Acid" in the REACH Database.* https://echa.europa.eu/information-on-chemicals/annex-iii-inventory/-/dislist/details/AIII-100.007.289

European Chemicals Agency. 2020d. *Entry for "2-Hydroxy Pyridine" in the REACH Database.* https://echa.europa.eu/substance-information/-/substanceinfo/100.005.019

European Chemicals Agency. 2020e. *Entry for "5-Methylhex-5-en-2-One" in the REACH Database.* https://echa.europa.eu/substance-information/-/substanceinfo/100.019.825

European Chemicals Agency. 2020f. *Entry for "Acetic Acid, Anhydride" in the REACH Database.* https://echa.europa.eu/substance-information/-/substanceinfo/100.003.241

European Chemicals Agency. 2020g. *Entry for "Benzoic Acid" in the REACH Database.* https://echa.europa.eu/substance-information/-/substanceinfo/100.000.562

European Chemicals Agency. 2020h. *Entry for "Butylated Hydroxy Toluene" in the REACH Database.* https://echa.europa.eu/substance-information/-/substanceinfo/100.004.439

European Chemicals Agency. 2020i. *Entry for "Decamethylcyclopentasiloxane" in the REACH Database.* https://echa.europa.eu/substance-information/-/substanceinfo/100.007.969

European Chemicals Agency. 2020j. *Entry for "OctamethylCyclotetrasiloxane" in the REACH Database.* https://echa.europa.eu/substance-information/-/substanceinfo/100.008.307

European Chemicals Agency. 2020k. *Entry for "HexamethylCyclotrisiloxane" in the REACH Database.* https://echa.europa.eu/brief-profile/-/briefprofile/100.007.970#collapseSeven

European Chemicals Agency. 2020l. *Entry for "Disiloxane Derivative" in the REACH Database.* https://echa.europa.eu/substance-information/-/substanceinfo/100.038.438

European Chemicals Agency. 2020m. *Entry for "Hydrochloric Acid" in the REACH Database.* https://echa.europa.eu/substance-information/-/substanceinfo/100.028.723

European Chemicals Agency. 2020n. *Entry for "Methyl Paraben" in the REACH Database.* https://echa.europa.eu/substance-information/-/substanceinfo/100.002.532

European Chemicals Agency. 2020o. *Entry for "TrimethylPhosphine" in the REACH Database.* https://echa.europa.eu/substance-information/-/substanceinfo/100.008.932

European Chemicals Agency. 2020p. *Entry for "Potassium Phosphate" in the REACH Database.* https://echa.europa.eu/substance-information/-/substanceinfo/100.029.006

European Chemicals Agency. 2020q. *Entry for "Sodium Hydroxide" in the REACH Database.* https://echa.europa.eu/substance-information/-/substanceinfo/100.013.805

European Chemicals Agency. 2020r. *Entry for "Thietane" in the REACH Database.* https://echa.europa.eu/substance-information/-/substanceinfo/100.005.469

Frickel, Scott, Sahra Gibbon, Jeff Howard, Joanna Kempner, Gwen Ottinger, and David Hess. 2010. "Undone Science: Charting Social Movement and Civil Society Challenges to Research Agenda Setting." *Science, Technology & Human Values* 35: 444–473.

Galison, Peter. 2008. "Removing Knowledge: The Logic of Modern Censorship." In *Agnotology: The Making and Unmaking of Ignorance,* edited by Robert Proctor and Londa Schiebinger, 37–54. Stanford: Stanford University Press.

Green Party. 2002a. "Painted Apple Moth." *Scoop Independent News,* July 4. https://www.scoop.co.nz/stories/PO0207/S00052.htm

Green Party. 2002b. "Government 'stacking' toxic spray advisory group." *Scoop Independent News,* November 7, 2002.

Green Party. 2002c. "Sutton Personal Attacks on Protestors Insensitive." *Scoop Independent News,* December 18, 2002.

Green Party. 2004. "Papers Prove MAF Tried to Delay Spray Report." *Scoop Independent News,* May 10, 2004. Accessed January 15, 2022. http://www.scoop.co.nz/stories/PA0405/S00161.htm

Goven, Joanna, Tom Kerns, Romeo Quijano, and Dell Wihongi. 2007. *Report of the March 2006 People's Inquiry Into the Impacts and Effects of Aerial Spraying Pesticide over Urban Areas of Auckland.* Auckland, NZ: Action Plan & Print. https://peoplesinquiry.wordpress.com/

Hales, Simon. 2004. "Precautionary Health Risk Assessment: Case Study of Biological Insecticides." *EcoHealth* 1: 399–403.

Hales, Simon, Virginia Baker, Kevin Dew, Losa Moata'ane, Jennifer Martin, Tim Rochford, David Slaney, and Alistair Woodward. 2004. *Assessment of the Potential Health Impacts of the 'Painted Apple Moth' Aerial Spraying Programme, Auckland.* Report for the New Zealand Ministry of Health.

Health Navigator New Zealand. n.d. "Asthma in Adults." https://www.healthnavigator.org.nz/health-a-z/a/asthma-adults/

Henderson, Julie, Annabelle Wilson, Samantha Meyer, John Coveney, Michael Calnan, Dean McCullum, Sue Lloyd, and Paul Ward. 2014. "The Role of the Media in Construction and Presentation of Food Risks." *Health, Risk & Society* 16(7–8): 615–630.

Hess, David. 2007. *Alternative Pathways in Science and Industry: Activism, Innovation, and the Environment in an Era of Globalization.* Cambridge, MA: MIT Press.

Jenkins, Jeffrey. 1992. "Environmental Toxicology and Chemistry Memo. Subject: B.t." Corvallis: Oregon State University Extension Service.

Kasperson, Roger and Jeanne Kasperson. 1996. "The Social Amplification and Attenuation of Risk." *ANNALS of the American Academy of Political and Social Science* 545: 95–105.

Kedgley, Sue. 2003. "Toxic ingredients of Painted Apple Moth spray tabled." Green Party Media Release, May 7. http://www.greens.org.nz/speeches/toxic-ingredients-painted-apple-moth-spray-tabled

Markowitz, Gerald and David Rosner. 2002. *Deceit and Denial: The Deadly Politics of Industrial Pollution.* Berkeley: University of California Press.

McGoey, Linsey. 2012. "Strategic Unknowns: Towards a Sociology of Ignorance." *Economy and Society* 41(1):1–16.

Michaels, David. 2008. *Doubt Is Their Product: How Industry's Assault on Science Threatens Your Health.* New York: Oxford University Press.

Ministry of Agriculture and Forestry (MAF). 2001a. "About the Btk Spray." *Scoop–Independent News,* October 25.

Ministry of Agriculture and Forestry (MAF). 2001b. "No Evidence of Health Problems From Btk Use." *Scoop–Independent News,* October 25.

Ministry of Agriculture and Forestry (MAF). 2002a. "Moth pest fight continues." *Scoop–Independent News,* March 5.

Ministry of Agriculture and Forestry (MAF). 2002b. "Cabinet decides on eradication for PAM." *Scoop–Independent News,* September 9.

Ministry of Agriculture and Forestry (MAF). 2002c. "GM risk from painted apple moth spray unlikely." *Scoop–Independent News,* October 10.

Moyers, Bill, Sherry Jones, Loren Berger, Jackson Frost, and Joseph Camp. 2002. *Trade Secrets: A Moyers Report.* Princeton, NJ: Films for the Humanities and Sciences.

New Zealand Herald. 2003. "Moth spraying will threaten growers' organic status." *New Zealand Herald,* August 13. https://www.nzherald.co.nz/nz/news/article.cfm?c_id=1&objectid=3517901

Novo Nordisk. 1991. *Material Safety Data Sheet for Foray 48B Flowable Concentrate.* Danbury, CT.

Office of the Ombudsman (Mel Smith). 2007. *Report of the Opinion of Ombudsman Mel Smith on Complaints Arising from Aerial Spraying of the Biological Insecticide Foray 48B on the Population of Parts of Auckland and Hamilton to Destroy Incursions of Painted*

Apple Moths, and Asian Gypsy Moths, Respectively. Wellington, NZ: Office of the Ombudsman.

Oreskes, Naomi and Erik Conway. 2010. *Merchants of Doubt.* New York: Bloomsbury Press.

Painted Apple Moth Community Advisory Group (PAM CAG). 2001. "Get Ruth-Less With the Painted Apple Moth." *Scoop Independent News,* December 17. https://www.scoop.co.nz/stories/PO0112/S00077.htm

Painted Apple Moth Community Advisory Group (PAM CAG). 2002a. "Halt to Aerial Spray Programme Called for." *Scoop Independent News,* June 24. https://www.scoop.co.nz/stories/PO0206/S00151.htm

Painted Apple Moth Community Advisory Group (PAM CAG). 2002b. "Re: Moth Pest Spray Date Announced." *Scoop Independent News,* July 4. https://www.scoop.co.nz/stories/PO0207/S00055.htm

Painted Apple Moth Community Advisory Group (PAM CAG). 2002c. "Community Advisory Group Harder to Get Rid of." *Scoop Independent News,* November 8. https://www.scoop.co.nz/stories/AK0211/S00030.htm

Painted Apple Moth Community Advisory Group (PAM CAG). 2003a. "4th damning report on Aerial Spray health effects." *Scoop Independent News,* April 2. https://www.scoop.co.nz/stories/AK0304/S00010.htm

Painted Apple Moth Community Advisory Group (PAM CAG). 2003b. "Painted Apple Moth Community Advisory Group." *Scoop Independent News,* June 23. https://www.scoop.co.nz/stories/PO0306/S00125.htm

Painted Apple Moth Community Coalition. 2002. "Evidence of adverse impacts becomes overwhelming." *Scoop Independent News,* December 18. https://www.scoop.co.nz/stories/AK0212/S00098.htm

Patnaik, Pradyot. 2007. *A Comprehensive Guide to the Hazardous Properties of Chemical Substances.* Hoboken, NJ: John Wiley & Sons.

Phillips, David. 2003. "Commentary to the MoH on Report Titled: 'Draft Interim Report of the Community-based Health & Incident Monitoring of the Aerial Spray Programme January–December 2002.'"

Proctor, Robert. 2008. "Agnotology: A Missing Term to Describe the Cultural Production of Ignorance (and Its Study)." In *Agnotology: The Making and Unmaking of Ignorance,* edited by Robert Proctor and Linda Schiebinger, 1–36. Stanford, CA: Stanford University Press.

Rayner, Steve. 2012. "Uncomfortable Knowledge: The Social Construction of Ignorance in Science and Environmental Policy Discourses." *Economy and Society* 41(1): 107–125.

Sanahuga, Georgina, Raviraj Banakar, Richard Twyman, Teresa Capell, and Paul Christou. 2011. "*Bacillus thuringiensis*: A Century of Research, Development and Commercial Applications." *Plant Biotechnology Journal* 9: 283–300.

Solleveld, Michelle, Anouk Schrantee, Nicolaas Puts, Liesbeth Reneman, and Paul Lucassen. 2017. "Age-Dependent, Lasting Effects of Methylphenidate on the GABAergic System of ADHD Patients." *Neuroimage Clinical* 15: 812–818.

Smith, Barbara Ellen. 1981. "Black Lung: The Social Production of Disease." *International Journal of Health Services* 11(3): 343–359.

Sutton, Jim. 2002. "All New Zealanders need to work together for effective biosecurity." *Beehive NZ Government Press Releases,* December 17, 2002. Accessed January 21, 2015. http://www.beehive.govt.nz/release/all-new-zealanders-need-work-together-effective-biosecurity

Swadener, Carrie. 1994. "*Bacillus thuringiensis* (B.T.)." *Journal of Pesticide Reform* 14(3): 13–20.

Tayabali, Azam and Verner Seligy. 2000. "Human Cell Exposure Assays of *Bacillus thuringiensis* Commercial Insecticides: Production of Bacillus cereus-Like Cytolytic Effects from Outgrowth of Spores." *Environmental Health Perspectives* 108(10): 919–930.

Valadares de Amorim, Giovana, Beatrixe Whittome, Benjamin Shore, and David Levin. 2001. "Identification of *Bacillus thuriengiensis* subsp. *kurstaki* Strain HD1-Like Bacteria from Environmental and Human Sample after Aerial Spraying of Victoria, BC, Canada with Foray 48B." *Applied and Environmental Microbiology* 67(3): 1035–1043.

Valent Biosciences. 2009. *Biological Insecticide Foray 48B Label.*

Vallée, Manuel. 2018. "Pharmaceuticalization through Government Funding Activities: ADHD in New Zealand." In *Global Perspectives on ADHD: Social Dimensions of Diagnosis and Treatment in Sixteen Countries*, edited by Meredith Bergey, Angela Filipe, Peter Conrad, and Ilina Singh, 288–309. Baltimore, MD: John Hopkins University Press.

van Netten, Chris, Kay Teschke, Victor Leung, Yat Chow, and Karen Bartlett. 2000. "The measurement of volatile constituents in Foray 48B, an insecticide prepared from *Bacillus thuringiensis var. kurstaki*." *The Science of the Total Environment* 263(1–3): 155–160.

Vizcaya, David, Maria Mirabelli, Josep-Maria Antó, Ramon Orriols, Felip Burgos, Lourdes Arjona, and Jan-Paul Zock. 2011. "A Workforce-Based Study of Occupational Exposures and Asthma Symptoms in Cleaning Works." *Occupational and Environmental Medicine* 68(12): 914–919.

Walsh, Frances. 2003. "Wipe Out." *Metro Magazine* March: 44–51.

Watts, Meriel. 2003. *Painted Apple Moth Eradication Programme: Health Risk and Effects.* https://www.semanticscholar.org/paper/PAINTED-APPLE-MOTH-ERADICATION-PROGRAMME-%3A-HEALTH-Watts/758c4f98b73b1e619 0d09facf8639763ee68a805

West Aucklanders against Spraying (WASP). 2002a. "BTK Spraying In Middle of Busy Shopping Centre." *Scoop Independent News,* June 17. https://www.scoop.co.nz/stories/AK0206/S00051.htm

West Aucklanders against Spraying (WASP). 2002b. "Painted Apple Moth Aerial Spraying." *Scoop Independent News,* July 4. https://www.scoop.co.nz/stories/PO0207/S00053.htm

9 The Mediating Role of Cultural Context

As discussed in the last two chapters, prior to and during the PAM spraying operation government officials undertook significant ideological work to allay citizen concerns about the Foray 48B pesticide. This included portraying the pesticide as harmless to humans, refusing to produce science that might produce uncomfortable knowledge, and neutralizing knowledge that could undermine the government's narrative about pesticide safety. Illuminating these activities helps explain why opposition to the spraying operation was so slow to develop.

However, it needs to be emphasized that these activities did not occur in a cultural vacuum. Citizen susceptibility to such communication activities will be mediated by their pre-existing knowledge and understanding of the world, which could provide them with the knowledge and inclination to be critical and effective citizens, or an ignorance and acculturation that makes them easier to manipulate. One particularly important mediator is their attitude towards pesticides. If citizens are predisposed to view pesticides as harmful substances, they are more likely to resist and question government messaging that suggests pesticides are harmless. Conversely, if they are inclined to view pesticides as relatively benign, they will be more predisposed to accept government information that suggests a particular pesticide is harmless. Another important mediator is their conceptualization of the state. If they believe the state prioritizes the health and well-being of its citizens, they are less likely to question government messaging. Then again, if they view the state as prioritizing industry profitability over public health, they will be more likely to be suspicious of government communications. A third mediator is their understanding of the politics of knowledge and ignorance production. If citizens do not understand the processes through which government messaging is produced and how the process is carefully crafted to both manage public opinion and maintain social control, citizens are less likely to be critical of such communications and more apt to accept them at face value. So, to better understand why opposition spread so slowly, we need to consider the cultural lens through which citizens interpreted the government's communication activities.

Another important part of the equation is the education system, which can provide students with the knowledge that will enable them to be effective

DOI: 10.4324/9780429426414-10

citizens, or, through a failure to adequately cover key issues, produce an ignorance that will make them easier to manipulate. For example, if the education system fails to educate all students about pesticides and toxicants, the citizenry, as a whole, is much less likely to understand how harmful these substances can be and much more likely to view pesticides as being relatively benign.

Although it is important to consider the entire education system, universities deserve particular attention as they are supposed to be the critical conscience of society. This is particularly true in New Zealand, whose Education Act explicitly singles out universities as being the critic and conscience of society. As the critic and conscience of society, universities have a responsibility to provide students with the knowledge, orientation, and skills that will enable them to be effective citizens, which includes giving them the skills and inclination to effectively assess and critique mass communications, question harmful cultural norms and practices, effectively oppose harmful government and corporate activities, and develop fairer and more sustainable ways of organizing society. Conversely, failing to provide this cultural equipment makes it harder for students to be the effective citizens they could and should be. Moreover, it makes it easier for governments and corporations to manipulate them.

This chapter explores five aspects of the cultural context that increased the effectiveness of the government's communication activities: (1) society's widespread acceptance of pesticide use; (2) factors contributing to that acceptance, including pesticide ignorance, an acquiescence to toxicity, and the deficient education provided about toxicants; (3) the deficient instruction universities provide regarding the government's role in capitalist political economies; (4) the universities' failure to adequately educate all students about the politics surrounding the production of knowledge and ignorance; and (5) the society-wide inclination to give academics free reign regarding university research and teaching.

The Cultural Acceptance of Pesticide Use

An important aspect of the cultural landscape in First World countries (such as Australia, France, Canada, New Zealand, and the United States) is the mass acceptance of pesticide use.

A global indicator of this acceptance is the amount of pesticides used, with global use surpassing six billion pounds in 2012 (Atwood & Paisley-Jones 2017). Another indicator is that pesticides continue to be used even though every year environmental scientists find evidence that commonly consumed produce (such as apples, grapes, peaches, strawberries, and cherries (see Table 9.1 for the Environmental Working Group's full 2022 list)) have concerning pesticide levels (Curl et al. 2003; EWG 2018, 2019, 2021; Fenske et al. 2002; Lu et al. 2010; MacIntosh et al. 2001; Morgan et al. 2005; Ripley et al. 2000; USDA 2013; Ye et al. 2015). The point is further amplified by the

Table 9.1 EWG's 2021 Dirty Dozen

(1) Strawberries	(7) Cherries
(2) Spinach	(8) Peaches
(3) Kale, collard, mustard greens	(9) Pears
(4) Nectarines	(10) Bell and hot peppers
(5) Apples	(11) Celery
(6) Grapes	(12) Tomatoes

Source: Environmental Working Group (2021).

fact these food items are favored with children, who are the most vulnerable to toxicants.

The continued purchase of pesticide-contaminated foods is a clear indicator the masses have accepted pesticide use as a standard agricultural practice. Some might disagree, pointing to the organic industry's growth as an indicator of consumer dissatisfaction with conventional agriculture. While it is undeniable that a growing organics sector suggests *some* consumers are dissatisfied with conventional agriculture, the organics industry only represents a tiny sliver of the overall agriculture market in Western democracies. For example, organic sales only accounted for 4% of 2021 food sales in the United States (USDA 2021). This is even truer in New Zealand, where organics only represented 2.5% of the country's food and beverage market in 2018 and where it was only 0.1% in 1997, two years prior to the start of the PAM eradication operation (Jones & Mowatt 2016; OEANZ 2018).

Another indicator of mass acceptance is that pesticide use has persisted even though environmental health researchers have been relating pesticide exposure to a growing list of health problems, including asthma, anemia, immune suppression, fertility problems, cognitive impairment, and cancers (see Table 9.2 for a full list). The point is further underscored by the fact researchers have found children are particularly vulnerable to pesticides' effects, because: (1) they consume larger amounts of food on a per kilogram basis; (2) their diets are less varied and they are more likely to eat foods with higher pesticide residues; (3) they have a weaker ability to metabolize and eliminate chemicals; (4) their developing systems (including neurological, digestive, and immunological) are more sensitive to pesticides' harmful impacts; (5) pesticides can disrupt developing systems that can lead to life-long repercussions; and (6) pesticides have been associated with a wide range of children's health problems, including delayed neurodevelopment and neurobehavioral deficits, lowered intelligence, altered growth, decreased lung function, and certain cancers (Guillette et al. 1998; Hyland & Laribi 2017; Marks et al. 2010; Raanan et al. 2016; Roberts & Karr 2012).

Although New Zealand likes to portray itself as being more environmentally sensitive than other countries, there is much evidence to suggest that it too had widespread acceptance of pesticides and that this was the case *prior to* the PAM eradication operation. First, there is the large number of

Table 9.2 Medical Problems Related to Pesticide Exposure

Strong Evidence	Good Evidence	Good Evidence
Arrhythmias	Abnormal sperm	Lymphoma
Contact dermatitis	Adult-onset leukemias	Menstrual disorders
Peripheral neuropathy	Aplastic anemia	Multiple myeloma
Reduced fertility – males	Asthma	Mycosis fungoides
	Bone cancer	Pancreatic cancer
	Brain cancer – children	Parkinson's disease
	Childhood leukemias	Photosensitivity
	Cognitive impairment	Psychiatric disturbances – disorientation, anxiety/ depression, emotional lability, and psychosis
	Decreased coordination	Reduced fertility – females
	Fetotoxicity	Renal cancer
	Genito-urinary malformations	Seizures
	Hormonal changes	Skin cancer
	Immune suppression	Testicular cancer
	Low birth weight	

Source: CHE (2019).

pesticides that were approved for sale in the country. By the late 1980s there were 280–300 pesticide active ingredients registered for use in New Zealand, with 700–900 formulations being used (MacIntyre et al. 1989; Watts 1994). Additionally, pesticide use grew significantly over time, with over 3,500 tons sold in 2000 (Manktelow et al. 2005). Third, pesticide use has consistently led to the pollution of water wells and streams (Close & Flintoft 2004; Close et al. 2021; Close & Skinner 2012; Gaw et al. 2008; Hageman et al. 2019). Fourth, New Zealand also has numerous food items with elevated pesticide levels, including grapes, strawberries, nectarines, and oranges (for the full list, please see Table 9.3) (Ceres Organics 2018; White 2014). Fifth, New Zealanders have demonstrated a tendency to continue using harmful pesticides long after they have been banned in comparable countries. Examples include DDT, chlorpyrifos, methyl bromide, diazinon, atrazine, glyphosate, 2,4-D, dimethoate, 1080, methamidophos, and 2,4,5-T, to name but a few (Castle 2020; Johnsen 2018; Wall 2018).

The New Zealanders' cultural acceptance of pesticides meant they viewed pesticide use as an acceptable way of addressing ecological problems. By extension, this would have inclined them to view pesticide spraying as a normal and acceptable response to eradicate an unwanted foreign species, particularly if the species was presented as a triple biosecurity threat to the country. In turn, this would have, naturally, softened their view of MAF's spraying operation, making them less inclined to criticize it and contributing to the slow spread of opposition.

Table 9.3 NZ's Dirty Dozen (2014)

(1) Grapes	(7) Spring onion
(2) Celery	(8) Lemons
(3) Bok/Pak choi	(9) Wheat
(4) Nectarines	(10) Cucumber
(5) Oranges	(11) Pears
(6) Strawberries	(12) Broccoli

Source: White (2014).

Factors Contributing to Pesticide Acceptance

While it is important to note the high level of pesticide acceptance, it is also important to unpack that acceptance and to identify the factors that enable and feed it. These include pesticide ignorance, an acquiescence to toxicity, deficiencies in the education curriculum, and the deeply held belief that academics should have unrestrained freedom regarding research and teaching about technology.

Pesticide Ignorance

The general acceptance of pesticides and the limited opposition to urban aerial pesticide spraying suggest the general populace does not understand (1) that engineered pesticides are composed of harmful synthetic chemicals (including neurotoxicants in most cases); (2) the processes through which pesticides kill their targets; and (3) how those same processes can harm other ecosystem inhabitants, including humans. While a case could be made that farmers, due to their first-hand experience and training, have a partial understanding of what pesticides consist of and how harmful they can be to humans, such partial knowledge is typically absent among urban dwellers, who are far less likely to have first-hand experience with pesticides (Salameh et al. 2004).

Moreover, an argument can be made that New Zealand urbanites, as a whole, suffer from greater pesticide ignorance than is the case in other Western democracies, as many seem to mistakenly believe that organic food is unnecessary in New Zealand. Specifically, in trying to explain the slow growth of New Zealand's organics industry, Jones and Mowatt (2016) found evidence suggesting citizens had bought into their government's "100% Pure and Natural" greenwashing campaign, which has them believing that pesticide use is very low in New Zealand and that, therefore, purchasing organic food is unnecessary. Ironically, this marketing campaign wasn't designed to fool New Zealanders but rather to fool international markets into thinking New Zealand products were healthier than they actually are (ibid). While current New Zealanders suffer from a profound pesticide ignorance, we can surmise this was even more so prior to and during the PAM spraying operation, as it was an era where even less media attention (on TV, in newspapers, and on social media) was devoted towards pesticide problems. This is also supported by the country's low rate of organic sales in 1997, which was only 0.1% of total food sales (Jones & Mowatt 2016).

The population's ignorance had important implications for the PAM spraying operation. First, citizens were less able to identify falsehoods in their government's communications about pesticide safety, which made them more susceptible to believing the proposed spraying operation would be completely safe. If a citizenry has an adequate understanding of what ingredients are in pesticides and how those ingredients can harm people, they will, presumably, have higher standards that need to be met in order to establish a pesticide's safety. Possessing such knowledge would have made them more critical of government attempts to portray a pesticide as being completely harmless to humans. Additionally, having adequate pesticide knowledge would have made the citizenry more resistant to appeals to ignorance, such as those New Zealand officials used in their communication campaign. In turn, having a public that was more critical about government claims would have increased the citizens' likelihood to oppose the PAM pesticide spraying operation.

The effects of pesticide ignorance would be trivial issues if the world was free of pesticides and other toxicants with harmful ingredients. This, however, is not the world we inhabit, not even "Clean and Green" New Zealand, which continues to produce and use copious amounts of harmful synthetic pesticides and other toxicants.

It is true that a major spraying operation was carried out in East Auckland in 1996–1997, that hundreds of people reported health effects from the spraying, and that these events could have informed West Aucklanders about what they were in for with the PAM spraying operation. However, Auckland is a large and sprawling city where social problems tend to be compartmentalized in the areas where those problems occur. This would have hindered the transmission of information between neighborhoods. Speaking to this point are the remarks from one West Auckland activist, who explained why it was that West Aucklanders didn't oppose the 1996–1997 pesticide spraying in East Auckland:

> Those of us in West Auckland just carried on our daily lives, concerned but safe in the knowledge it would not affect us, as it was too far away. Also, many were not aware of all the adverse health effects suffered by the East Aucklanders. Auckland is a big place - very spread out, with localised communities.

Given that this view was held by an activist who was inclined to be aware of health issues, it suggests that typical citizens in West Auckland would have been even less aware of the East Auckland spraying operation.

An Acquiescence to Toxicity

Although pesticide ignorance helps to explain the high level of support citizens gave to MAF, it appears there is an additional contributing factor, based on the fact that ten months into the spraying operation 51% of spray zone residents still supported the MAF's pesticide spraying operation, with only 13% registering dissent (Office of the Ombudsman 2007). That population's high rate of support is striking because they repeatedly experienced the sprayings and would have had first-hand knowledge about its potential impact on human health, local ecology, animals, and their homes. While they might have lacked a formal education about the topic, one imagines their first-hand knowledge and their proximity to other impacted residents would have led far more of them to oppose the spraying operation.

To make sense of this outcome I draw on Woodhouse and Howard's (2009) "acquiescence to toxicity" concept, which refers to the tendency of First World citizens to believe that high levels of toxicants (such as pesticides) are the necessary cost of living in an affluent consumer society. Woodhouse and Howard (2009) argue "most citizen-consumers, government officials, business executives, and chemists have made this key assumption, which has made good public policy about chemicals almost impossible" (p. 53). Although they maintain that economic elites and government agencies play a key role in perpetuating high levels of toxicity, Woodhouse and Howard also emphasize the complicity of citizen-consumers. This complicity manifests itself when citizens continue to purchase toxic materials, fail to question either the lack of healthier alternatives or the politicians that perpetuate the status quo, and fail to adequately mobilize against new toxic threats.

There is much to suggest New Zealanders have had an acquiescence to pesticides, including their heavy use of pesticides over time, allowing their government to subsidize pesticide use up until the mid-1980s, continuing to use pesticides and eat pesticide-laden foods despite the growing knowledge of health problems associated with pesticides, and the country being an international laggard when it comes to banning the most harmful pesticides, such as DDT, chlorpyrifos, methyl bromide, diazinon, atrazine, glyphosate, 2,4-D, dimethoate, 1080, methamidophos, and 2,4,5-T, to name but a few (Castle 2020; Johnsen 2018; Wall 2018). These all signal a strong reliance on pesticides and an inability to wean themselves off of this technology.

The acquiescence to toxicity concept illuminates another aspect of the PAM story. Even if West Aucklanders were opposed to the PAM pesticide spraying, they would have been predisposed, at least initially, to see it as a necessary cost for eradicating the painted apple Moth.

A Deficient Education about Toxicants

As Woodhouse and Howard (2009) point out, education systems contribute significantly to the ignorance most Western citizens have towards toxicants. The reason is that they fail to adequately inform students about the tens of thousands of toxicants (including pesticides) being used throughout society or the potential human and ecological harmfulness caused by those toxicants.

Although education systems are failing to properly educate students about toxicants, this is particularly true of universities, which, as a group, have failed to adequately educate the general population about the problem. This is not to suggest universities fail to offer *any* courses on toxicology, as many do. However, most toxicology courses tend to focus on pharmacological toxicity, rather than the toxic products we are exposed to in our home, work, educational and recreational environments. Moreover, when toxicological courses are on the books, they are offered as electives, instead of core curriculum that everyone has to take. This means the vast majority of students taking the courses are those with a natural affinity for the topic or those who need it to satisfy a graduation requirement. The consequence is that the vast majority of students pass through university without learning what pesticides are composed of, how they function, or how they harm human and other ecosystem inhabitants. In sum, they graduate without the knowledge that would help them navigate a society where pesticides, and toxicants more generally, are an everyday reality and threat.

This problem is quite germane to New Zealand. To this day, the country's university system, along with the rest of the country's education system, fails to provide all students with an adequate literacy about pesticides and other toxicants. Such a literacy would include developing an awareness of (1) which synthetic chemicals are being used in society; (2) how they are produced; (3) the financial, ecological, and public health costs associated with their production; (4) who benefits from producing and using those toxicants and how; (5) what impact those chemicals have on people exposed to them; (6) what impact their use has on ecosystems; (7) how people can protect themselves and their communities from toxicants in their living, work, and recreational environments; and (8) the cultural and political tools citizens can be used to eliminate pesticides and other toxicants considered harmful to humans and ecosystems.

This situation would not be alarming in a country that has low pesticide usage. However, such has not been the case in Aotearoa New Zealand, which has a history of using copious amounts of pesticides, of being a laggard when it comes to banning particularly harmful pesticides, and of having had the audacity, on three separate occasions, to carry out aerial pesticide spraying operations over densely populated urban neighborhoods.

One factor contributing to the situation is that most universities have continued to cling to the neoliberal idea that a university's main educational mission is to provide a marketplace of ideas, where students are encouraged to build an à la carte education. This is in contrast to providing a comprehensive education that provides students with the tools they will need to effectively address the problems in their society. Unfortunately, continuing down à la carte "marketplace of ideas" pathway is a form of educational and moral negligence. While a literacy about toxicants would not have been necessary in a pre-industrial society, we now live in a society where every day we are repeatedly exposed to toxicants, through the products we apply to our skins, the foods we eat, the water we drink, and the air we breathe. Thus, an adequate knowledge about toxicants is a pre-requisite for safely navigating a world replete with potential toxic dangers and higher education's failure to provide that knowledge reflects a commitment to a curriculum that is not fit for purpose and is a profound moral failing vis-à-vis the next generation.

Universities could alleviate the problem by instituting a general education graduation requirement that would acquaint students with pesticides and/or toxicants more generally. An example would be the "environmental awareness" general education courses that have been instituted at San Francisco State University, the University of Vermont, and the public universities in the state of Minnesota. Unfortunately, such examples are rare exceptions. In a recent survey I found that less than 20 out of 550 public universities in the United States require any sort of "environmental awareness" graduation requirement. A similar problem has existed in "Clean and Green" New Zealand, where, as of 2022, no university had instituted such a graduation requirement. This includes the University of Auckland, which trumpets itself as being the country's "leading university."

The absence of "environmental awareness" graduation requirements means that, for decades now, New Zealand universities have been failing to provide students with the knowledge that could help them protect themselves, their families, and their larger communities in pesticide-laden communities. While it is true many people have a general sense they should avoid exposing themselves to pesticides, most do not properly understand what are in pesticides, how they work, why they are harmful to humans, the health problems they are linked to, or how to pressure their government officials to tighten regulations around pesticide production and use. In turn, this pesticide ignorance has made it less likely that opposition to pesticides would spread among the masses.

Imparting a Deficient Conception of the State

Another problem with university curricula is the deficient conception of the state it imparts to most students. As Pellow (2009) remarks, a particularly enduring academic view of the state is that it is pluralist in orientation, positing the state's role as being a neutral arbitrator between the different social forces competing for state resources, government influence, and power. In the pluralist view interest groups and associations work to influence state policymaking, thereby giving citizens multiple channels for voicing concerns.

Unfortunately, however, such a conception is deficient for most Western democracies. While a pluralist conception is certainly more apt than an autocratic one in such nations, many social scientists, particularly conflict theorists, argue, a pluralist conception of the state is deficient as it obscures the way government agencies are captured by economic forces (Pellow 2009; Schnaiberg & Gould 1994). Marx's (Tucker 1978) position on the issue is that the state is merely industry's executive, whose purpose is to assist industry's efforts to maximize profit accumulation. A more recent and more nuanced version of this position is Schnaiberg and Gould's (1994) "treadmill of production" model, which views governments as being embedded in a system geared towards economic growth, which, as a result, leads them to pursue activities that help accelerate environmental destruction. In this model, while citizens of democratic nations certainly have more potential influence than would be the case in autocratic nations, that potential influence is heavily curbed by the significant influence exerted by economic forces, who have successfully lobbied individual politicians and, even more importantly, have shaped the paradigms through which decisions get made (Carolan 2017).

The treadmill of production model captures well the New Zealand situation, where there has been a long history of governments facilitating environmental destruction to advance industry interests. Examples include land development policies that caused large-scale deforestation, subsidy schemes that led to the drainage of wetlands, political support for energy projects that despoiled lakes and rivers, permitting environmentally destructive mining, and creating subsidy schemes that intensified the use of synthetic pesticides and fertilizers (Bührs 1993; Dann 2002).

However, despite the New Zealand government's long history of enabling environmental destruction, the curricula of New Zealand universities still fail to impart an adequate conceptualization of the state to all students. While universities might offer a handful of courses that convey a more critical conceptualization of the state (including sociology courses), such courses are always electives and have limited enrolments (typically under 120), which means they are only taken by a tiny fraction of students. This matters because if universities are only imparting a critical conception of the state to a small fraction of students, the remainder are being taught to adopt a pluralist conception of the state, where government agencies are seen as neutral arbiters between the different conflicting groups. Moreover, by failing to see government agencies as being slanted towards industry interests, the masses are less likely to hold a critical stance vis-à-vis government policies, activities, and communications.

The implication for the PAM spraying operation is that operating with a deficient conception of the state decreased the citizens' likelihood of developing a critical stance vis-à-vis either the government or their communications campaign about the pesticide. This, in turn, facilitated government efforts to portray the pesticide as harmless and slowed the spread of opposition.

Educational Deficiencies Regarding the Production of Knowledge and Ignorance

Another deficiency with university curricula is the failure to educate all students about the politics of knowledge, which includes knowing not just how the production and communication of scientific knowledge is shaped by political, economic, and cultural forces, but also the way ignorance is carefully cultivated in order to manage public opinion and manufacture consent.

In recent decades social scientists have shown that the production of scientific knowledge is much more politicized than most people realize, as they have illuminated that powerful forces (including corporations and governments) shape what topics get researched, how they get researched, and what can be communicated about that research (Cozzens & Woodhouse 1995; Frickel & Moore 2006; Kleinman 2001; Schwartz 1996). Additionally, the literature has revealed that the proponents of pesticides, toxicants, and other consumer products have tended to portray their products in a rosier light than is usually warranted (Davis 2007; Déplaude 2015; Proctor 2008). This includes portraying their products as completely safe even when no safety data has been carried out, or when safety data is still being compiled, or when safety data showing harm has been suppressed (Abramson 2005; Angell 2005; Davis 2010; Scott 2004; Vogel 2012). The latter is a particularly common problem with the pharmaceutical industry, where companies have raised to an art form the practice of concealing and obscuring data showing harm, and have been aided by captured government regulators, and medical professionals that have been trained to administer pharmaceuticals without fully understanding the medications' health ramifications (Abramson 2005; Angell 2005; Scott 2004). A similar situation exists with pesticides, as there is a long history of governments approving pesticide products under the assumption they are safe, only to later ban them when they have been proven to be harmful to ecosystems and humans (Collins & Johnston 2005; Fletcher 2018; Taylor 1997; UNEP 1996; Wall 2018).

Social scientists have also shed important light on the politics of ignorance, identifying how ignorance is not just an absence of knowledge but rather is a strategic resource that corporate and government forces can carefully cultivate in order to better control public opinion on a range of issues (Frickel & Vincent 2011; McGoey 2012; Oreskes & Conway 2010; Proctor 2008). Beyond portraying products in a falsely positive light, the proponents of harmful technologies also work to suppress information that is inconvenient to their agenda (Steve Rayner 2012). This process has also been extensively studied by social scientists, who have documented numerous strategies through which social forces suppress the production of uncomfortable knowledge. This includes preventing the funding of research on harmful substances, incentivizing researchers to focus on non-threatening questions, or intimidating researchers who are thinking of pursuing threatening questions (Frickel et al. 2010; Hess 2007; Markowitz & Rosner 2002; Proctor 2008).

As well, social scientists have documented the tactics through which social forces seek to neutralize uncomfortable knowledge that has been produced, which includes suppressing its circulation by intimidating its producers, discrediting such knowledge and/or its producers, downplaying its importance, or distracting audiences away from it (Déplaude 2015; Markowitz & Rosner 2002; Proctor 2008; Rayner 2012).

Although social scientists have produced knowledge regarding the politics of knowledge and ignorance, universities in New Zealand and other Western democracies have done a poor job of imparting this knowledge to all students. To be clear, I am not arguing they fail to provide any courses on the issue, as some universities do provide such content via a handful of courses (such as *Sociology of Knowledge*, *Sociology of Science*, and *Science and Technology Studies*). Rather, I am arguing such institutions are the exception rather than the rule as such courses are completely absent from the course listings of most colleges and universities in Western democracies. Moreover, when such courses are offered, they are electives, which means the courses are only taken by the small number of students who opt to take them. The consequence is that the vast majority of students graduate without understanding how knowledge production is highly mediated by powerful forces, how ignorance benefits those in power, how ignorance is actively produced, or how they can protect themselves from processes aimed at maintaining their ignorance.

These are major deficiencies to have in an information age, where the average citizen is routinely bombarded with information, misinformation, and disinformation. Moreover, such deficiencies increased the effectiveness of the New Zealand government's communication campaign, as the population's widespread ignorance about the politics of knowledge and ignorance made it less likely they would take a critical stance vis-à-vis the government's claims that the pesticide was harmless, or any other part of the government's million-dollar communications campaign.

Conversely, if citizens had been educated about the politics of knowledge and ignorance, they would have been equipped to identify strategies that were used to perpetuate ignorance, such as those the New Zealand government used to neutralize uncomfortable knowledge (including suppression, omission, dismissal, denial, and distraction). If a public is aware of tactics to suppress and neutralize uncomfortable knowledge, they are more likely to recognize when such tactics are deployed against them and will be able to more effectively resist them.

Root Causes of Educational Deficiencies

The deficiencies with the university curriculum were significant and to shed deeper light on the topic, this section discusses two factors that contributed to them, which include the neoliberal conception of universities as marketplaces of ideas and the citizenry's deference to academics when it comes to university research and teaching.

The Neoliberal Conception of Universities as Marketplaces of Ideas

The educational deficiencies can be partially traced to the ruling neoliberal paradigm that views universities as marketplaces of ideas, and which has led students to an à la carte approach to their education, where they often pick a range of different courses that have seemingly little connection to each other.

While universities should be providing a breadth of choices for students to pick from, they should also be providing the knowledge students will need to effectively function in the type of society they will inhabit. If people live in a society that readily uses thousands of synthetic chemicals on food and other consumer products and most of those chemicals have yet to be adequately tested for safety, then it is reasonable to expect universities will educate students about those substances, so that they can adequately protect themselves, their families, and their larger communities. Additionally, if students are to be effective citizens in democratic societies, it is reasonable to expect that universities will educate students about the way knowledge and ignorance are socially produced, and which interests government agents will favor in a capitalist political economy.

Although universities have succeeded in providing students a cornucopia of courses, they have failed to ensure that every student emerges with an education that is fit for living in a modern world. This includes having an education that will enable them to effectively resist tactics of mass manipulation. Not only would having that knowledge immunize students against mass manipulation, it would also make them more effective at opposing corporate and government depredations, which would enable them to build healthier ecosystems and to produce a healthier population. One can't help but conclude that the university education currently provided is not fit for the purpose of building healthy, sustainable, and resilient communities. On the other hand, the knowledge deficiencies being produced do make it easier for governments to manipulate and control their citizens.

The Public's Deference to the Authority of Academics

Woodhouse and Howard's (2009) work provides additional insight about the deficiencies, as they point to the fact local communities rarely have any say about the type of research and teaching that is conducted in universities. For instance, they argue the public rarely weighs in about the type of chemistry research that is pursued or how it is pursued. Nor does the public have much, if any, say about what is taught about toxic products. In turn, this has led universities to reproduce curriculum that is geared towards protecting the industrial status-quo, rather than to prioritize curriculum that would help reduce the production of industrial waste, which despoils ecosystems and puts people in harm's way. This is manifested by the fact the vast majority of university chemistry departments continue to teach harmful "brown" chemistry, instead of the more benign green chemistry that was developed in the late 1990s. Out of the 1,200 tertiary institutions in North America, only eight have comprehensively incorporated green chemistry principles into their undergraduate curriculum (Vallée 2016). Moreover, only 38 have managed to add *any* green chemistry course to their curriculum (Vallée 2016).

Furthermore, no university program requires its graduate students (from which the future industry leaders will emerge) to take any toxicology courses (Woodhouse & Howard 2009).

Taking the analysis one step further, one reason the public has little to no say is that university research and teaching is often opaque to the larger public, which effectively prevents citizens from scrutinizing what is taking place in the academy. Additionally, Woodhouse and Howard (2009) trace the public's lack of participation to their delegation of authority to academics, which is based on the deeply engrained and widespread belief that academic scientists, engineers, and universities should have unrestrained freedom regarding research and teaching about technology. They also argue this "excessive academic freedom for scientists is rooted in a naïve, idealized view of science as selfless, independent truth seeking that more or less automatically improves the human condition" (2009: 49), which contrasts with a view of scientists as regular human beings, who are self-interested and who will pursue endeavors that will interest them and build their careers.

A potential solution to this issue is Michael Crow and William Dabar's (2018) model of the new research university, where universities act as knowledge hubs that work collaboratively with communities to develop knowledge that will be useful to the community, as opposed to producers of esoteric knowledge that may or may not be useful to local communities. This approach would provide local communities with greater transparency about the process through which knowledge is produced and transmitted, which would help them more effectively guide research and teaching towards community needs, which could include knowledge about pesticides and other toxicants, the government's inclination to favor industry interests in capitalist political economies, and the strategies and tactics governments and other social forces deploy to manipulate and control citizens.

Summary

This chapter started from the premise that government communications do not occur in a cultural vacuum, as the people exposed to those communications have a pre-existing ideological apparatus that will mediate how they respond to it. I then proceeded to identify key elements of the cultural context within which the PAM spraying operation took place, which included the widespread cultural acceptance of pesticide use that predisposed citizens to support the pesticide spraying operation. Additionally, this chapter discussed factors that fed the cultural acceptance, which included pesticide ignorance, an acquiescence to toxicity, and the universities' failure to adequately educate all students about pesticides and toxicants, in general. Another key cultural element was the deficient instruction students receive about the state and about the politics surrounding the production of knowledge and ignorance, all of which also made it harder for citizens to protect themselves against the government's communication campaign. Moreover,

I related these issues to the dominant neoliberal conceptualization of higher education and the public's deference to the authority of academics.

The main point of this chapter has been to underscore how cultural context can mediate the effectiveness of government communication campaigns and to illuminate key mediating elements of that context. Beyond helping to explain the PAM case, the analysis should be helpful for understanding the effectiveness of government communication campaigns associated with other urban aerial pesticide spraying operations, not to mention those associated with pesticide spraying of parks, sports fields, golf courses, roads, and agricultural crops.

References

Abramson, John. 2005. *The Broken Promise of American Medicine*. New York: Harper Books.

Angell, Marcia. 2005. *The Truth About the Drug Companies: How They Deceive Us and What to Do About it*. New York: Random House Trade Paperbacks.

Atwood, Donald and Claire Paisley-Jones. 2017. *Pesticides Industry Sales and Usage: 2008–2012 Market Estimates*. Washington, DC: US Environmental Protection Agency.

Bührs, Ton. 1993. "The Role of the State: From 'State Vandalism' Towards a 'Market-led' Environment?" In *Environmental Policy in New Zealand: The Politics of Clean and Green?* edited by Ton Bührs and Robert Bartlett, 90–112. Auckland, NZ: Oxford University Press.

Carolan, Michael S. 2017. *Society and the Environment: Pragmatic Solutions to Ecological Issues*. Boulder, CO: Westview Press.

Castle, Belinda. 2020. "Pesticides in fruit and vege." *Consumer.org*. https://www.consumer.org.nz/articles/pesticides-in-fruit-and-vege

Ceres Organics. 2018. *Dirty Dozen: The most pesticide-laden produce [plus Clean 15 list]*. https://ceres.co.nz/blog/dirty-dozen-the-most-pesticide-laden-produce-plus-clean-15-list/

Close, Murray E. and Mark Flintoft. 2004. "National Survey of Pesticides in Groundwater in New Zealand – 2002." *New Zealand Journal of Marine and Freshwater Research* 38(2): 289–299.

Close, Murray E., Bronwyn Humphries, and Grant Northcutt. 2021. "Outcomes of the First Combined National Survey of Pesticides and Emerging Organic Contaminants (EOCs) in Groundwater in New Zealand 2018." *Science of the Total Environment* 754(1): online. https://doi.org/10.1016/j.scitotenv.2020.142005

Close, Murray E. and A. Skinner. 2012. "Sixth National Survey of Pesticides in Groundwater in New Zealand." *New Zealand Journal of Marine and Freshwater Research* 46(4): 443–457.

Collaborative on Health and the Environment (CHE). 2019. "Pesticides." *Toxicants and Disease Database*. https://www.healthandenvironment.org/our-work/toxicant-and-disease-database/?showcategory=&showdisease=&showcontaminant=2380&showcas=&showkeyword=

Collins, Simon and Martin Johnston. 2005. "NZ Using Pesticides Barred in America" *New Zealand Herald*. https://www.nzherald.co.nz/nz/news/article.cfm?c_id=1&objectid=10006400

Cozzens, Susan and Ed Woodhouse. 1995. "Science, Government and the Politics of Knowledge." In *The Handbook of Science and Technology Studies*, edited by Sheila Jasanoff, Gerald Markle, James Petersen, and Trevor Pinch, 533–553. London: Sage Publications.

Crow, Michael M. and William B. Dabars. 2018. *Designing the New American University*. Baltimore, MD: John Hopkins University Press.

Curl, Cynthia L., Richard A. Fenske, and Kai Elgethun. 2003. "Organophosphorus Pesticide Exposure of Urban and Suburban Preschool Children with Organic and Conventional Diets." *Environmental Health Perspectives* 111(3): 377–382.

Dann, Christine. 2002. "Losing Ground? Environmental Problems and Prospects at the Beginning of the Twenty-First Century." In *Environmental Histories of New Zealand*, edited by Eric Pawson and Tom Brooking, 275–287. Auckland, NZ: Oxford University Press.

Davis, Devra. 2007. *The Secret History of the War on Cancer*. New York: Basic Books.

Davis, Devra. 2010. Disconnect: The Truth about Cell Phone Radiation, What Industry Has Done to Hide It and How to Protect Your Family. New York: Dutton Press.

Déplaude, Marc-Olivier. 2015. "Minimising Dietary Risk: The French Association of Salt Producers and the Manufacturing of Ignorance." *Health Risk & Society* 17(2): 168–183.

Environmental Working Group (EWG). 2018. *Dirty Dozen: EWG's 2018 Shopper's Guide to Pesticides in Produce*. Washington, DC: Environmental Working Group.

Environmental Working Group (EWG). 2019. *Dirty Dozen: EWG's 2019 Shopper's Guide to Pesticides in Produce*. Washington, DC: Environmental Working Group.

Environmental Working Group (EWG). 2021. *Dirty Dozen – EWG's 2021 Shopper's Guide to Pesticides in Produce*. https://www.ewg.org/foodnews/dirty-dozen.php

Fenske, Richard A., Golan Kedan, Chensheng Lu, Jennifer A. Fisker-Andersen, and Cynthia L. Curl. 2002. "Assessment of Organophosphorous Pesticide Exposures in the Diets of Preschool Children in Washington State." *Journal of Exposure Analysis and Environmental Epidemiology* 12: 21–28.

Fletcher, Raquel. 2018. "Québec Tightens Rules on Pesticides." *Global News*, February 19.

Frickel, Scott and Kelly Moore. 2006. "Prospects and Challenges for a New Political Sociology of Science." In *The New Political Sociology of Science*, edited by Scott Frickel and Kelly Moore, 3–25. Madison: The University of Wisconsin Press.

Frickel, Scott, Sahra Gibbon, Jeff Howard, Joanna Kempner, Gwen Ottinger, and David Hess. 2010. "Undone Science: Charting Social Movement and Civil Society Challenges to Research Agenda Setting." *Science, Technology & Human Values* 35: 444–473.

Frickel, Scott and Bess Vincent. 2011. "Hurricane Katrina, Contamination, and the Unintended Organization of Ignorance." *Technology in Society* 29: 181–188.

Gaw, Sally, Murray E. Close, and Mark J. Flintoft. 2008. "Fifth National Survey of Pesticides in Groundwater in New Zealand." *New Zealand Journal of Marine and Freshwater Research* 42(4): 397–407.

Guillette, Elizabeth A., Maria Mercedes Meza, Maria Guadalupe Aquilar, Alma Delia Soto, and Idalia Enedina. 1998. "An Anthropological Approach to the Evaluation of Preschool Children Exposed to Pesticides in Mexico." *Environmental Health Perspectives* 106(6): 347–353.

Hageman, Kimberly J., Christopher Aebig, Kim Hoang Luong, Sarit L. Kaserzon, Charles S. Wong, Tim Reeks, Michelle Greenwood, Samuel Macaulay, and

Christoph D. Matthaei. 2019. "Current-use Pesticides in New Zealand Streams: Comparing Results from Grab Samples and Three Types of Passive Sampler." *Environmental Pollution* 254(A): 112973.

Hess, David. 2007. *Alternative Pathways in Science and Industry: Activism, Innovation, and the Environment in an Era of Globalization.* Cambridge, MA: MIT Press.

Hyland, Carly and Ouahiba Laribi. 2017. "Review of Take-Home Pesticide Exposure Pathway in Children Living in Agricultural Areas." *Environmental Research* 156: 559–570.

Johnsen, Meriana. 2018. "Controversial Chemicals not on New Safety Review List." *Radio New Zealand,* October 16, 2018. https://www.rnz.co.nz/news/national/368744/controversial-chemicals-not-on-new-safety-review-list

Jones, Geoffrey and Simon Mowatt. 2016. "National Image as a Competitive Disadvantage: The Case of the New Zealand Organic Food Industry." *Business History* 58(8): 1262–1288.

Kleinman, Daniel. 2001. "Systemic Influences: Some Effects of the World of Commerce on University Science." In *Degrees of compromise: Industrial Interests and Academic Values,* edited by Jennifer Croissant and Sal Restivo, 225–239. Albany, NY: SUNY Press.

Lu, Chensheng, Frank J. Schenck, Melanie A. Pearson, and Jon W. Wong. 2010. "Assessing Children's Dietary Pesticide Exposure: Direct Measures of Pesticide Residues in 24-Hr Duplicate Food Samples." *Environmental Health Perspectives* 118(11): 1625–1630.

MacIntosh, David L., Caroline W. Kabiru, and P. Barry Ryan. 2001. "Longitudinal Investigation of Dietary Exposure to Selected Pesticides." *Environmental Health Perspectives* 109(2): 145–150.

MacIntyre, Angus A., Nicholas Allison, and David R. Penman. 1989. *Pesticides: Issues and Options for New Zealand.* Wellington, New Zealand: Ministry for the Environment.

Manktelow, David, P. Stevens, James Walker, S. Gurnsey, N. Park, J. Zabkiewicz, David Teulon, and A. Rahman. 2005. *Trends in Pesticide Use in New Zealand 2004.* Report to the Ministry for the Environment. Project No. SMF4193.78p.

Markowitz, Gerald and David Rosner. 2002. *Deceit and Denial: The Deadly Politics of Industrial Pollution.* Berkeley: University of California Press.

Marks, Amy R., Kim Harley, Asa Bradman, Katherine Kogut, Dana Boyd Barr, Caroline Johnson, Norma Calderon, and Brenda Eshkenazi. 2010. "Organophosphate Pesticide Exposure and Attention in Young Mexican-American Children: The CHAMACOS Study." *Environmental Health Perspectives* 118(12): 1768–1774.

McGoey, Linsey. 2012. "Strategic Unknowns: Towards a Sociology of Ignorance." *Economy and Society* 41(1): 1–16.

Morgan, Marsha K., Linda S. Sheldon, Carry W. Croghan, Paul A. Jones, Gary L. Robertson, Jane C. Chuang, Nancy K. Wilson, and Christopher W. Lyu. 2005. "Exposures of Preschool Children to Chlorpyrifos and Its Degradation Product 3,5,6-Trichloro-2-Pyridinol in Their Everyday Environments." *Journal of Exposures Analysis and Environmental Epidemiology* 15: 297–309.

Office of the Ombudsman. 2007. *Report of the Opinion of Ombudsman Mel Smith on Complaints Arising from Aerial Spraying of the Biological Insecticide Foray 48B on the Population of Parts of Auckland and Hamilton to Destroy Incursions of Painted Apple Moths, and Asian Gypsy Moths, Respectively.* Wellington, NZ: Office of the Ombudsman.

Oreskes, Naomi and Erik Conway. 2010. *Merchants of Doubt*. New York: Bloomsbury Press.

Organic Exporters Association of New Zealand (OEANZ). 2018. "Organics in New Zealand." https://www.organictradenz.com/organics-in-new-zealand.html

Pellow, David. 2009. "The State and Policy: Imperialism, Exclusion, and Ecological Violence in State Policy." In *Twenty Lessons in Environmental Sociology*, edited by K. Gould and T. Lewis, 47–58. New York: Oxford University Press.

Proctor, Robert. 2008. "Agnotology: A Missing Term to Describe the Cultural Production of Ignorance (and Its Study)." In *Agnotology: The Making and Unmaking of Ignorance*, edited by Robert Proctor and Linda Schiebinger, 1–36. Stanford, CA: Stanford University Press.

Raanan, Rachel, John R. Balmes, Kim G. Harley, Robert B. Gunier, Sheryl Magzamen, Asa Bradman, and Brenda Eskenazi. 2016. "Decreased Lung Function in 7-Year-Old Children with Early-Life Organophosphate Exposure." *Thorax* 71: 148–153.

Rayner, Steve. 2012. "Uncomfortable Knowledge: The Social Construction of Ignorance in Science and Environmental Policy Discourses." *Economy and Society* 41(1): 107–125.

Ripley, Brian D., Linda I. Lissemore, Pamela D. Leishman, Mary Anne Denomme, and Leonard Ritter. 2000. "Pesticide Residues on Fruits and Vegetables from Ontario, Canada, 1991–1995." *Journal of AOAC International* 83(1): 196–213.

Roberts, James R., Catherine J. Karr, and Council On Environmental Health. 2012. "Pesticide Exposure in Children." *Pediatrics* 130(6): e1765–e1788.

Salameh, Pascale R., Isabelle Baldi, Patrick Brochard, and Bernadette Abi Saleh. 2004. "Pesticides in Lebanon: A Knowledge, Attitude, and Practice Study." *Environmental Research* 94: 1–6.

Schnaiberg, Allan and Kenneth Alan Gould. 1994. *Environment and Society: The Enduring Conflict*. New York: St. Martin's Press.

Schwartz, Charles. 1996. "Political Structuring of the Institutions of Science." In *Naked Science: Anthropological Inquiry into Boundaries, Power, and Knowledge*, edited by Laura Nader, 148–159. New York: Routledge Press.

Scott, Catherine D. 2004. *Selling Sickness*. San Francisco, CA: Kanopy film.

Taylor, Rowan, Ian Smith, Peter Cochrane, Brigit Stephenson, and Nicci Gibbs. 1997. *The State of New Zealand's Environment 1997*. Wellington, NZ: Ministry for the Environment. ISBN 0-478-09000-5.

Tucker, Robert C. 1978. *The Marx-Engels Reader*. New York: Norton.

United Nations Environmental Programme. 1996. Rotterdam Convention – Operation of the Prior Informed Consent Procedure for Banned or Severely Restricted Chemicals Decision Guidance Documents 2,4,5-T and its salts and esters. Joint FAO/UNEP Programme for the Operation of Prior Informed Consent. http://www.pic.int/Portals/5/DGDs/DGD_2,4,5-T_EN.pdf

United States Department of Agriculture (USDA). 2013. *Pesticide Data Program Annual Summary Reports*.

United States Department of Agriculture – Economic Research Service (USDA). 2021. "Organic Agriculture." https://www.ers.usda.gov/topics/natural-resources-environment/organic-agriculture/

Vallée, Manuel. 2016. "Obstacles to Curriculum Greening: The Case of Green Chemistry." In *The Contribution of Social Sciences to Sustainable Development at*

Universities, edited by W. Leal Filho and M. Zint, 245–258. Cham, Switzerland: Springer International Publishing.

Vogel, Sarah. 2012. *Is It Safe? BPA and the Struggle to Define the Safety of Chemicals.* Berkeley: University of California Press.

Wall, Tony. 2018. "It's Banned in Other Countries but New Zealand Is Using More Toxic Methyl Bromide than Ever." *Stuff.co.nz.* June 4. Accessed August 30, 2020. https://www.stuff.co.nz/national/103690904/its-banned-in-other-countries-but-new-zealand-is-using-more-toxic-methyl-bromide-than-ever

Watts, Meriel. 1994. *The Poisoning of New Zealand.* Auckland, NZ: Auckland Institute of Technology Press.

White, Alison. 2014. "The 'dirty dozen' – latest update." *Organics NZ,* January 1, 2014. https://organicnz.org.nz/magazine-articles/dirty-dozen-latest-update/

Woodhouse, Edward and Jeff Howard. 2009. "Stealthy Killers and Governing Mentalities: Chemicals in Consumer Products." In *Killer Commodities: Public Health and the Corporate Production of Harm*, edited by Merrill Singer and Hans Baer, 35–60. Plymouth, UK: Altamira Press.

Ye, Ming, Jeremy Beach, Jonathan W. Martin, and Ambikaipakan Senthilselvan. 2015. "Associations between Dietary Factors and Urinary Concentrations of Organophosphate and Pyrethroid Metabolites in a Canadian General Population." *International Journal of Hygiene and Environmental Health* 218: 616–626.

Conclusion

Over the last 29 years governments in New Zealand, Canada, and the United States have carried out more than 20 aerial spraying operations over densely populated residential areas, including in Charlotte, North Carolina (1992, 1998, 2008); Spokane, WA (1993); Auckland, New Zealand (1996–1997 and 2002–2004); Hamilton, New Zealand (2003–2004); Victoria, British Columbia (1998); Seattle, Washington (2000, 2016); Toronto, Ontario (2007, 2008, 2013, 2017, 2019, and 2020); Burnaby, British Columbia (2022); Langley, British Columbia (2022); and Surrey, British Columbia (2015, 2019, 2020, and 2022).

What these cases have in common is that citizens were informed that a foreign species had established itself; that the way to contain the incursion was to conduct an aerial pesticide spraying operation; and that while the ingredients could not be revealed to the community, they had nothing to worry about because pesticide was harmless to humans.

Auckland's 2002–2004 operation was not the first time an aerial pesticide spraying operation was conducted in a densely populated residential area, nor was it the last, as over a dozen have occurred since then. However, what made the Auckland case so remarkable was its unprecedented scope, duration, and the number of people it exposed to pesticide spraying, which has never been surpassed. Over the course of 29 months the New Zealand Ministry of Agriculture and Forestry (MAF) administered aerial pesticide spraying on 60 days, over an area that reached 10,632 hectares (26,272 acres) at its peak and that repeatedly exposed 200,000 people to the spraying (Goven et al. 2007; Office of the Ombudsman 2007). Making this case even more intriguing is that it took place in a country with a strong environmental reputation.

Illuminating the Social Processes that Led to the Spraying Operation

One of the book's objectives has been to illuminate the social processes that led to such an unprecedented pesticide spraying operation. First, this book examined the social processes that contributed to PAM's arrival and spread

DOI: 10.4324/9780429426414-11

throughout West Auckland. An important contributing factor was the global trade network's expansion, which increased the number of sites from which foreign species could arrive. Another factor was increased global trade, which generated more opportunities for foreign species to travel to New Zealand. Additionally, technological improvements sped up transit times, which increased the likelihood a foreign species would survive the trip from its native locale to the new one. Still another contributing factor was the country's unwillingness to adequately fund biosecurity monitoring. Beyond factors that increased the chances a foreign species could arrive in New Zealand, MAF completely botched its initial response to PAM's arrival, allowing it to spread considerably over the course of three years.

A second key issue was the government's decision to eradicate the moth. As mentioned in Chapter 3, this response was not a given as the country had experienced the arrival of 19,000+ foreign species in the 160 years prior to PAM's arrival and only a small fraction of those species were ever targeted for eradication. I argued that the government's desire to eradicate PAM was shaped by several factors, including: (1) a dominant worldview where ecosystems are perceived and treated as a resource base; (2) the country's long history of struggling with foreign species; (3) its strong economic reliance on primary industry; (4) the fact PAM could feed on pine trees, which raised the theoretical possibility that the moth might reduce industry profitability; (5) a political economy geared towards maximizing economic growth; and (6) the industrialization of primary industries, whose reduction of biodiversity increased industry vulnerability to invasive species.

A third important issue was the government's decision to seek eradication via pesticide spraying, which was also not a foregone conclusion as there were other potential strategies they could have deployed (including mating disruption technologies, such as using pheromone traps and sterilized moths). To make sense of why they pursued pesticides in this instance, the decision has to be embedded within its larger cultural context. I argued that a key aspect of this context is the "synthetic age" (Foster 1999) we live in, where we are predisposed to using technological solutions to address our problems, including ecological problems. Along these lines, New Zealand manifested a strong affinity for using pesticides to solve its ecological problems, as underscored by New Zealand governments: (1) approving a large number of pesticides since the 1950s; (2) implementing subsidy schemes to encourage pesticide use; (3) repeatedly using pesticides for national operations to control pests; and (4) being laggards in the banning of particularly harmful pesticides. The other factor contributing to the aerial spraying was MAF's inept initial handling of PAM, which allowed the moth to spread well beyond its original point of arrival, which contributed to the perception the country was facing a biosecurity crisis. In turn, plantation forestry executives, in particular, leveraged that crisis perception to pressure MAF into escalating its response with an aerial spraying operation.

A key issue in this story was the local community's response to the spraying operation. While there was a small group of dedicated people who worked fiercely and tirelessly to oppose the spraying operation, that opposition did not stop MAF from carrying out its operation or expanding it. In part, this outcome can be traced to the fact government agents, prior to the PAM operation, altered the Biosecurity Act to give MAF a legal loophole that would enable them to pursue eradication actions without having to comply with local concerns. Although the legal loophole was crucial, government efforts were also considerably aided by the slow spread of public opposition to the spraying. The slow spread of opposition can be partially attributed to fear-inducing government communications that framed the moth as a triple biosecurity threat: a threat to the economy, to native ecology, and to public health. Fearmongering of this sort is a particularly powerful strategy as humans are hard-wired to respond to fear.

While they were stoking fear about PAM, New Zealand officials also pursued actions to allay citizen concerns about the pesticide, which included proposing a small and limited spraying operation (six to eight sprayings over four months, with a spray zone of 300 hectares) at the start, while also portraying the pesticide and its ingredients as harmless to humans. Moreover, when they expanded the operation, they did so incrementally, which made it more difficult to notice and oppose. Government officials also allayed concerns by managing "uncomfortable knowledge" (Rayner 2012) about the pesticide. In doing so their strategies included refusing to produce knowledge about the human health impact of the sprayings, and, when forced to produce such knowledge, impeding the production and dissemination of such knowledge. Government agents also neutralized existing knowledge through suppression, omission, dismissal, denial, downplaying, and diversion. As well, they hampered potential sources of uncomfortable knowledge, whose information could have undermined support for the government's agenda.

Lastly, this book considered how the effectiveness of the government's communication campaign was enhanced by the ideological context. Part of that context included the widespread acceptance of pesticide use, which was itself fed by pesticide ignorance, an acquiescence to toxicity, and the education system's failure to adequately educate all citizens about pesticides and other toxicants. Another key aspect of the ideological context was the deficient education citizens receive regarding the role governments play in capitalist political economies and the politics surrounding the production of knowledge and ignorance. Additionally, the educational deficiencies are themselves fed by the widely accepted beliefs that universities should be organized as marketplaces of ideas and the public should defer to the wisdom of academics regarding the research and teaching taking place in the ivory tower.

Insights Generated about Urban Aerial Spraying Operations

Although the Auckland case was remarkable for its scope and duration, it did trigger substantial condemnation, in the form of the People's Inquiry, the report that emerged from that inquiry (Goven et al. 2007), and the

Ombudsman's 2007 report, all of which contributed to delegitimizing the use of such spraying operations in New Zealand. Remarkably, however, despite the condemnation produced in New Zealand, aerial pesticide spraying has continued to be used in cities across North America, where at least 12 aerial spraying operations have been administered since the end of New Zealand's PAM operation. Consequently, while this book's primary objective was to illuminate the Auckland case, another objective was to use this case to generate knowledge and insights about why and how urban aerial pesticide spraying operations come about, with the hope that this knowledge could be used to reduce their likelihood in the future.

The first insight is that pesticide spraying operations are the outgrowth of numerous social processes, including those leading to (1) increased opportunities for foreign species to travel from one locale to another; (2) failing to detect the foreign species when they arrive; (3) failing to contain the species once detected; (4) deciding to eradicate the species; (5) deciding to use pesticides to do so; (6) convincing the public of the need to eradicate the species; and (7) allaying citizen concerns about the pesticide.

Another insight generated by this analysis is that the arrival of a foreign species is not a random occurrence but rather is socially produced, in that it is strongly mediated by human behavior, social systems, and social policy. For instance, the willingness to trade with countries that harbor different species opens the possibility for foreign species to migrate from that country. Additionally, increasing trade with those countries grants those species more opportunities to migrate to the new locale. Moreover, adopting technology that accelerates transportation speeds will increase the likelihood that foreign species will survive passage to new locales. Beyond factors that increase the volume of new arrivals, a foreign species' ability to establish in a new country will be mediated by the biosecurity agreements and protocols established between the trading countries, as well as the importing country's biosecurity monitoring apparatus, whose effectiveness will be mediated by national funding policy and, ultimately, national value systems. Another important element is the political economy, which will shape every step of the process. For instance, a capitalist political economy that is fixated on increasing trade at all costs will encourage expanding trade routes and increasing the volume of trade, while discouraging biosecurity protocols and/or monitoring systems that could stifle trade.

Thus, if we are serious about reducing foreign species incursions, our approach to the problem should go beyond a knee-jerk reaction that focuses simplistically on eliminating the targeted species. Instead, we should adopt an expansive perspective, which understands that the likelihood of incursions is mediated by "socio-biological networks" (Jay & Morad 2006). As well, we should target our efforts to addressing the human behaviors, policies, and systems that are contributing to the problem. Not only would such an approach be wiser, it would, ultimately, be cheaper and would produce less suffering (Mack et al. 2000).

A third insight is that the decision to eradicate a foreign species is mediated by a complex set of social factors, including the dominant ecological world-view operating in that country, the country's history of struggling with foreign species, its economic reliance on primary industries, the degree to which those sectors have been industrialized, and the perceived threat of the species in question. Additionally, the perceived threat is shaped by framing activities pursued by government agents, whose effectiveness will be mediated by the university education citizens receive about ecosystems, governments, and critical thinking.

One way to lessen the drive to eradicate is to reduce the perceived and actual vulnerability that primary industries feel towards foreign species, which could be done by implementing practices that regenerate, rather than destroy, ecosystem biodiversity. One such practice would be for operators to move away from obsessively selecting genetic strains that grow the biggest or fastest and to instead emphasize strains that are more resistant to diseases, insects, as well as variations in moisture and temperature. As well, operators could move away from monocropping, as poly-cropping techniques also increase resilience vis-à-vis environmental disturbances. A third beneficial shift would be to move away from synthetic pesticides and towards integrated pest management systems, which control pests by, among other things, having crop variation, using companion planting, as well as encouraging insect biodiversity and healthy bird populations. Of course, adopting these techniques would be facilitated by shifting the political economy towards valuing sustainability, biodiversity, and resilience, instead of a single-minded focus on maximizing economic growth.

Fourth, a set of insights emerged around the biosecurity concept. One insight is that "biosecurity" is quite subjective, as what is considered a biosecurity threat will vary across time, space, and even between groups within a particular society. Thus, the use of the term has to be carefully contextualized. The analysis also revealed the fearmongering strategies that government officials deploy to frame some species as threats, in an effort to drum up support for spraying activities. Thus, those studying pesticide spraying activities should pay particular attention to the use of such controlling processes. A third insight is that, as emphasized throughout the text, defining something as a biosecurity threat has important social implications, as some stand to benefit from framing a species as a threat, while others can be harmed by the ripple effects of those framing activities. This underscores that we should carefully analyze who does and does not benefit from such framing activities. Another insight is that, as underscored by New Zealand's 1993 Biosecurity Act, the biosecurity concept can be codified into law, which can significantly enhance governmental powers to deal with species they consider undesirable. Still another insight is that the biosecurity concept can be weaponized, in that government agents can use the concept to abusively impose harmful practices on some segments of society and to shut down dissent. This development is particularly chilling for the protection of human rights and democracy.

Moreover, while the focus here was on pesticide spraying, one imagines how similar calls to "protect" national security could open the door to the imposition of other unjust and harmful technologies and/or practices.

Fifth, this book elucidates that decisions to use pesticides also have to be contextualized, as they too are shaped by a complex set of social forces, which include primary industries, environmental groups, local community groups, and government agents, who can deploy an array of strategies to allay citizen concerns. Furthermore, the effectiveness of government strategies will be mediated by the degree to which a population accepts pesticide use, as well as their knowledge/ignorance about pesticides, the role governments play in capitalist political economies, and the politics associated with the production of knowledge and ignorance. Furthermore, the knowledge/ignorance about these issues is itself shaped by university education.

Another point underscored by this book is that governments play a central role in the emergence of aerial pesticide spraying operations. This includes their role in producing foreign species incursions, which occurs as a result failing to (1) require shipping companies to implement protocols that will minimize the arrival of new species; (2) adequately fund the border monitoring aimed at detecting the arrival of new species; (3) invest in the leadership required to effectively contain a newly discovered foreign species without putting people in harm's way; and (4) require primary industry operators to implement biodiversity-enhancing practices, which would give the country any extra layer of biosecurity against invasive species. Additionally, government agents play a central role in building public support for the pesticide spraying, using fearmongering tactics to stimulate support for eradicating the foreign species, and using numerous strategies and tactics to allay citizen health concerns about the pesticide in question.

Lastly, this book emphasizes the importance of the educational context. Citizens who suffer from key educational deficiencies (including about ecosystems, toxicants, and the inclination of governments in capitalist political economies) will have more difficulty defending themselves against government strategies deployed to build support for pesticide use. For instance, if citizens are not educated about how ecosystems work, they are more likely to fall for diagnoses that simplistically problematize the foreign species and ignore the contextual elements that contributed to the species' arrival. Similarly, if students are not educated about how toxicants can harm humans, they will be more apt to believe soothing government reassurances that a pesticide is harmless to humans. These problems are exacerbated when universities also fail to educate students about government affinities for prioritizing industrial interests, as such students will be more apt to see government claims as unbiased and legitimate. Additionally, it is easier for government officials to get away with making unsupported and/or fallacious arguments when universities fail to adequately educate students about how to critically assess claims and identify rhetorical fallacies. Regarding the latter, if students have not been adequately trained to defend themselves against these rhetorical dark arts, they are unlikely to recognize when government agents are trying to manipulate them by appealing to ignorance and/or appealing to their fears, or deploying some other rhetorical fallacy.

All of this underscores that in order to effectively defend themselves against the propaganda used to support pesticide spraying, citizens require an education that is fit for the purpose of living in a capitalist political economy. This includes adequately educating students about ecosystems, toxicants, the tendency for government in capitalist political economies to champion industry interests, the politics around the production of knowledge and ignorance, the tools to critically assess claims, and how to recognize rhetorical fallacies. Having such an education will help them understand which questions they need to ask and what criticisms they need to make when confronted with government propaganda. For example, if government officials claim that a moth will negatively impact 95% of humans, as the New Zealand officials claimed, citizens should (A) be aware that governments in a capitalist political economy are invariably aligned with industry interests and will disseminate information that serves industry interests, which should reduce the reliability of their claims; (B) recognize the rhetorical effects associated with making appeals to fear and understand that such appeals are a powerful strategy to shut down critical thinking and manipulate the masses; (C) ask politicians how they know their claim to be true; (D) demand to see proof supporting their claims; and (E) when such proof fails to materialize, they should know the importance of heartily denouncing politicians for spreading unverifiable alarmist information.

When the citizenry is armed in this way, they will be far more effective at defending themselves against the tactics that governments deploy to support pesticide use, which will make them far more effective at opposing pesticide spraying. This pertains to spraying operations to eliminate foreign species as well as other pesticide uses. Importantly, universities should be leading the way in these efforts, as they have a responsibility to do more than train the next generation to be the rearguard of a vandal economy (Orr 1995). In particular, they have a responsibility to give their students the intellectual tools to effectively critique the status quo, as well as envision and produce a more just, sustainable, resilient, nurturing, and thriving world. Granted, transforming our educational institutions in this way will not be easy. Then again, the most worthwhile things in life are rarely easy to attain. Moreover, as Margaret Mead reminds us, when a small group of committed people sets themselves to a task, even the most difficult social transformation can be accomplished: "Never doubt that a small group of thoughtful and committed citizens can change the world. In fact, it is the only thing that ever has."

References

Foster, John Bellamy. 1999. *The Vulnerable Planet*. New York: Monthly Review Press.
Goven, Joanna, Tom Kerns, Romeo Quijano, and Dell Wihongi. 2007. *Report of the March 2006 People's Inquiry Into the Impacts and Effects of Aerial Spraying Pesticide over*

Urban Areas of Auckland. Auckland, NZ: Action Plan & Print. https://peoplesin-quiry.wordpress.com/

Jay, Mairi and Munir Morad. 2006. "The Socioeconomic Dimensions of Biosecurity: The New Zealand Experience." *International Journal of Environmental Studies* 63(3): 293–302.

Mack, Richard N., Daniel Simberloff, W. Mark Lonsdale, Harry Evans, Michael Clout, and Fakhri A. Bazzaz. 2000. "Biotic Invasions: Causes, Epidemiology, Global Consequences, and Control." *Ecological Applications* 10(3): 689–710.

Office of the Ombudsman. 2007. *Report of the Opinion of Ombudsman Mel Smith on Complaints Arising from Aerial Spraying of the Biological Insecticide Foray 48B on the Population of Parts of Auckland and Hamilton to Destroy Incursions of Painted Apple Moths, and Asian Gypsy Moths, Respectively.* Wellington, NZ: Office of the Ombudsman.

Orr, David W. 1995. "Reinventing Higher Education." In *Greening the College Curriculum: A Guide to Environmental Teaching in the Liberal Arts*, edited by Jonathan Collett, 8–23. Washington, DC: Island Press.

Rayner, Steve. 2012. "Uncomfortable Knowledge: The Social Construction of Ignorance in Science and Environmental Policy Discourses." *Economy and Society* 41(1): 107–125.

Index

Note: **Bold** page numbers refer to tables; *italic* page numbers refer to figures.